MINING, METALLURGY
AND THE
MEANING OF LIFE

A Book of Stories

Roger Sworder

MINING, METALLURGY

AND THE

MEANING OF LIFE

A Book of Stories

SOPHIA PERENNIS

SAN RAFAEL, CA

First edition, Quaker Hill Press, 1995
Second edition, Sophia Perennis, 2008
© Roger Sworder 2008

Series editor: James R. Wetmore

For information, address:
Sophia Perennis, P.O. Box 151011
San Rafael, CA 94915
sophiaperennis.com

Library of Congress Cataloging-in-Publication Data

Sworder, Roger.
Mining, metallurgy, and the meaning of life:
a book of stories / by Roger Sworder—1st American ed.

p. cm.
Originally published: Sydney, Australia:
Quakers Hill Press, 1995.
Includes bibliographical references and index.
ISBN 978-1-59731-085-7 (pbk: alk. paper)
1. Metallurgy—Religious aspects. 2. Mines and mineral resources
—Religious aspects. 3. Metallurgy in literature. I. Title.
TN673.S86 2008
669—dc22   2008022416

CONTENTS

Preface: *City of Gold*   *v*

1. *Origins*   *1*
   Mining and the Land
   Slave Mining and the Right to Work

2. *The Classical Tradition*   *21*
   Homer's Smith God
   Prometheus and Arachne
   Plato's Theory of Craft

3. *The Biblical Tradition*   *42*
   Greeks, Jews and Egyptians
   Genesis and Exodus
   Solomon and Daniel
   The New Testament

4. *The Medieval Synthesis*   *76*
   The Saintly Patronage of Mining and Metallurgy
   Precious Metals in Church Worship
   Christian Alchemy

5. *The Symbol of the Mine*   *106*
   Dragons
   Dwarfs

6. *The Desacralisation of Work*   *121*
   The Traditional Work Ethic
   The Protestant Work Ethic
   Blake and Wordsworth on Work and Nature

Epilogue: *The Nuclear Age*   *149*
Notes   *155*
Book References   *162*
Index   *167*

CITY OF GOLD

I AM WRITING THESE words in a nineteenth century house on the goldfields of Australia. I live in Bendigo, a small city in the state of Victoria, about one hundred miles inland from Melbourne, the state capital on the southern coast of the continent. At its peak in the 1880s Bendigo housed perhaps seventy-five thousand people. Here there came, from the 1850s onwards, thousands of miners from all over the world, whose work inaugurated an amazing local prosperity. No doubt most of the gold which they dug from this earth was sent back to England. But enough of the wealth stayed to pay for splendid public buildings and a main street which is the equal, I was told when I arrived, of the Champs Élysées in Paris.

The city was built with money from gold. The old lady who sold this house to us told us that when she was a young girl in Bendigo and wanted something from the shops, her mother would tell her to go out and find the gold to pay for it. This was easiest after rain when the tiny particles glittered in the sun. Australia is an antediluvian continent long unknown to the west. But in Bendigo at least, its soil has been thoroughly sifted. The very land beneath this house is a labyrinth of galleries. I write, I hope, in something of the spirit of those early citizens as I relate the stories from the Bible and elsewhere which transformed their understandings and gave meaning to their arduous lives.

For they were not quite the people we are. Most of those nineteenth century miners and their families lived in two different worlds at once, their own and also the world of the Old and New Testaments. They were not so insulated in their portion of the historical continuum as we have become in ours. They believed themselves close to an everliving centre, the Saviour who is present equally to every point on the arc of time. This land, too, was the scene of salvation, though it was very far from the Holy Land. They neither were, nor did they think themselves a mere aggregation of productive units as economic historians tend to think of them now. It is true that this period of Australia's history is easy to explain in economic terms. And in the terms of the twentieth century, the religious aspirations of those people may be described as a delusion, by which they concealed from themselves the real nature of their humanity. But their gifts to the city suggest much more, that their beliefs reflected a deeper level of self consciousness than we recognise now.

In our street there are three large churches and in the next there is a cathedral. Across the park from here is the nineteenth century School of Mines and Industry with its octagonal library decorated by a famous German designer. Here was housed the collection of books on all subjects, especially literature and philosophy, which was the nucleus of the Mechanics Institute. On this side of the street from the School of Mines is the furthest corner of the Conservatory gardens which run alongside most of the length of the main street and which border the park. In the middle of the park stands the red brick Gothic pile of Camp Hill Primary School, on the site of the early garrison in Bendigo. Next to it is the Georgian front of the Bendigo High School. The plan of the city bespeaks a great love of its children. Behind and above the two schools Camp Hill rises, and on its summit is placed a poppet head, one of the pylons raised over mining shafts. The outline of a poppet head, gold on a background of dark blue, is the symbol of the city.

If we stand on the platform at the top of this poppet head, from which the great wheel of the pulley has been removed, we can view the bowl or basin of land in the centre of which the city is set. To the south the horizon is ringed with hills, to the north it stretches away across a plain. This bowl of land was originally a forest which was cut down by the miners and used in their work. But this forest has long regrown and the city rises and falls to the lie of the land among trees so thickly clustered that it is hard to see the houses from this height. On the outskirts of the city to the south, on the very edge of the bush, is the local campus of the university. In its library are many of the volumes of that earlier collection which are still in use by students of literature, philosophy and history. We who work there are lineal descendants of the School of Mines.

I am most grateful to Dr John Webster, presently Chief Executive of the Institution of Engineers, Australia, and formerly Director of the University College in Bendigo during the time this book was written; and to Ray Evans of Western Mining Corporation who first proposed to me a book on mining, metallurgy, and religion; to Rev. Dr Evan Burge, John Bradley and Dr Findlay Johnson for their help with Hebraic and metallurgical matters; to Clive Faust, Maurice Nestor and Clifford Carrington for their ideas and books; to Madge Pinge, Dorothy Avery and Anne Tyndall for their help with apparatus mechanical and bibliographical; and to Peter Day of Quakers Hill Press who has greatly improved the manuscript by his suggestions. Needless to say, none of the above is committed to what follows.

Chapter One

ORIGINS

OF ALL THE CRAFTS and professions other than the priesthood, none has been more closely connected with the religious traditions of western peoples than mining and metallurgy. Not very long ago our ancestors would have found it incredible that people could not see the connections between mining, metallurgy and the sacred just as we now find it incredible that they could. What was a commonplace to the European mind for millennia has now become a matter of the deepest obscurity.

This is a matter of more than historical interest. It goes to the heart of how we think about work, about religion and about the relations between people and nature. First, there is the dignity of work as this was understood in earlier times. For our ancestors most forms of work were spiritual paths, disciplines which shaped those who engaged in them as powerfully as the rituals of church or temple. In many crafts and professions the stages by which the learner was inducted were initiations into substantial understandings of the spirit and of spiritual practice. Not to work was to forgo an essential means to spiritual development, without which one stood in immediate danger of damnation. To work, on the other hand, was to make oneself over to a practice which was regarded as much greater than the individual, through which alone the worker could be fulfilled. In the early chapters of this book we consider how the smith god of the Greeks and Romans made the world in his forge, how Moses made the tabernacle as God commanded, how mining was sanctified in the middle ages. The contemplation of such things restores to the miners and metallurgists of every succeeding age something of that splendour and pride in work which invested all the ancient crafts and professions. It raises such work far above the level of profit, product and wages, and confers upon the worker a share in divine powers.

That work developed the human spirit was once a commonplace. The loss of this belief has profoundly impoverished the lives of many western peoples over the last two or three centuries. Equally, it has impoverished our sense of religion and of the sacred. According to our ancestors a very large proportion of any human population is predisposed to the growing or rearing of plants and animals and to the manufacture of artifacts. Traditionally these forms of work were thought to repeat and extend the creative powers of God and nature. Those who engaged in them enjoyed a special insight into the processes of the divine creation. They knew from their own experience how the world proceeded from the invisible mind into manifestation. The loss of such understandings has left us with a notion of God to which we may assent, but which we can hardly feel. No longer a living presence realised daily in our work, the bare existence of God has become a matter of abstract argument. The withdrawal of the sacred sense from human work has diminished religion in many western societies. The several stages by which this withdrawal occurred is one of the major concerns of this book, especially in the later chapters.

This same withdrawal has also diminished our sense of the relations between the human and natural worlds. We tend now to oscillate between the view that nature is a perfect and untouchable power of which we should be in awe, and the view that it is a mere resource to be used up and discarded. In earlier times nature was understood quite differently. Working with the natural world helped it to bring to birth the many goods with which it was in labour. Work fulfilled not only the worker but nature itself. Even in quite recent times the spirits of the earth, the fairies and dwarfs, actively assisted farmers and miners in their work according to common belief. They rewarded diligence and thrift and punished lazy and indifferent workers. They represented in more or less human form the elemental creative energies of the natural world, and they revealed themselves precisely to the people who worked to release those energies. In the fifth chapter of the book we look at some of the stories about these elemental denizens of the mine and how they shaped the attitudes of miners to their work. And in the last chapter we consider how the belief in such creatures came to be lost and the consequences of that loss for our understanding of the natural world. The study of this throws a new light on the environmental debates of our own time, debates which do not begin with us but have occupied the western mind for millennia.

These are large and complex issues, much more complex than our present understandings of work, religion and nature would suggest. From the point of view of the theories advanced in this book, our own understandings of these things are narrow. There is no period of the western past in which they have been limited as they are now. It is not so much that our views differ from those of earlier times as that we have no views. Rather than having developed a different philosophy we may be said to have no philosophy. This book is a book of stories from many different places and times in the history of the west. The juxtaposition of these stories in a coherent sequence reveals a way of looking at work, nature and religion which is much more substantial than our own. For all that, we must begin from where we are, with an examination of some objections to mining and metallurgy which are often made today. And even here we will find that our concerns have long been anticipated, and that where we think we are most modern, we are often most traditional.

MINING AND THE LAND

The earliest denunciations of mining and metallurgy which have survived to us in the west come from the reigns of the Emperor Augustus and his immediate successors. Augustus inaugurated a new era in the history of ancient Rome after a long period of civil war. He established himself securely as the sole ruler of Rome and its empire, transformed its political institutions and brought peace. His court encouraged the arts, especially poetry, and the poets responded by presenting the new Augustan order as a return of the golden age. Augustus himself came from a banking family. But despite his banking origins Augustus did much to restore the traditional agrarian economy of Rome and Italy. The great armies of the civil wars were disbanded and put back on the land, and in this transformation the poets played the part of propagandists for the simple life.

In the very first years AD the poet Ovid wrote the *Metamorphoses*, and it is here that we find the first extended attack on the mining and working of metals. Ovid described in detail the four ages of humankind which he denominated in the traditional manner by the names of the four metals: gold, silver, bronze and iron. With the advent of the iron age, according to Ovid, the land which had been common to all was first marked out by boundary lines. Then for the first time humankind spread sails to the wind and insolently crossed the sea. And then, too, for the first time man began to dig into the very bowels of the earth where the wealth, which the creator had hidden away, lay buried among the shades of the dead. Then was harmful iron discovered, and gold more harmful even than iron. And from these discoveries came war and love of plunder and crime. Ovid's emphasis on the use of metals in war as a feature of the iron age may reflect the connection between the Augustan peace and the new golden age.

Meanwhile others wrote manuals of farming which were widely distributed. Rome itself was part of the land of Latium which was believed to have been governed in earliest times by King Saturn, before he was displaced by Jupiter from the throne of heaven. When Saturn ruled, humankind had lived in the golden age in perfect harmony and simplicity, provided with everything they needed by a bountiful nature which required no toil. To this age Ovid looked back as did many of the poets contemporary with him, and in this they furthered the political aims of Augustus as he worked to re-establish as much of the old Roman world as could survive.

There was at least a literary rediscovery of the rustic idyll, in which the real and simple wealth which farming produced was implicitly contrasted with the artificial wealth of commerce, mining and conquest. In this way Ovid's account of the discovery and use of metals takes its place in a much larger program of civil and economic restoration.

In Ovid's account of mining for metals two objections were implicit. First there is the implication that since the creator hid the metals deep in the earth, they ought to stay there. Mining is a contravention of the divine order. But this objection may be made to almost every human action; Ovid himself criticises sailing. It makes sense in this form only when we take into account the Saturnian idyll of the golden age when nature supported humankind without requiring any effort on our part. Secondly, Ovid pointed out that the metals were hidden where the spirits of the dead reside, and here the implication is that mining is a profanation of the infernal realm. This objection is rather more complex than at first appears. That mining has to do with the realm of the dead enables us to compare it with ancient views of agriculture. For here too it was believed that breaking though the surface of the earth to plough and sow brought humankind into contact with the infernal powers. For this reason and because it required an assault on mother earth, agriculture required a special sanction. This was provided by the cult of Demeter, the Roman Ceres, and her daughter Persephone or Proserpine. The story went that Hades or Pluto, the god of the dead, abducted Demeter's daughter and took her down to the underworld to be his wife. But Demeter prevailed upon the gods to compel Pluto to return her daughter. A compromise was reached which required the girl to spend some time each year in the underworld and the remainder with her mother in heaven. In this way the cycle of the agricultural year received divine authority; and Demeter taught humankind the art of ploughing. The celebration of the cult of Demeter, Persephone and Pluto was one of the most important in the ancient world.

The story and the cult of Demeter sanctified the practice of agriculture in the ancient world. It transformed the god of the departed from a figure of dread into one of the sources of agricultural wealth. In Greek, for example, the word for wealth was *plutos*. So when Ovid described the place where the creator hid the metals as the place of the departed spirits, he touched upon a theme which his readers could hardly have failed to associate with the cult of agriculture. The vital question is whether the practice of mining was sanctified by a cult of this same kind, which propitiated the powers of the earth in the infernal regions. There is no evidence of such a thing on anything like the scale of the Demeter cult. Nonetheless Ovid's reference to the realm of the departed would have evoked a complex of feelings in his audience. His hearers were accustomed to regard that realm both fearfully and with gratitude as the source of their prosperity. In this way Ovid's attack on mining brought the issue well within the religious sensibility of his time.

Ovid's objections to mining were quickly followed by the more extended attack of Pliny the Elder in his Natural History. Pliny was born soon after Ovid's death and he clearly knew and approved of Ovid's objections. This is how he began the books on metals at the end of his *Natural History:*

> Our topic now will be metals, and the actual resources employed to pay for commodities; resources diligently sought for in the bowels of the earth in a variety of ways. For in some places the earth is dug into for riches, when life demands gold, silver, silvergold and copper, and in other places for luxury, when gems and colours for tinting walls and beams are demanded, and in other places for rash valour, when the demand is for iron, which amid warfare and slaughter is even more prized than gold. We trace out all the fibres of the earth, and live above the hollows we have made in her, marvelling that occasionally she gapes open or begins to tremble, as if it were not possible that this may be an expression of the indignation of our holy parent! We penetrate her inner parts and seek for riches in the abode of the spirits of the departed, as though the part where we tread upon her were not sufficiently bounteous and fertile... The things that she has concealed and hidden underground, those that do not quickly come to birth, are the things that destroy us and drive us to the depths below; so that suddenly the mind soars aloft into the void and ponders what finally will be the end of draining her dry in all the ages, what will be the point to which avarice will penetrate. How innocent, how blissful, how luxurious life might be, if it coveted nothing from any source but the surface of the earth, and, to speak briefly, nothing but what lies ready to her hand!

Here again we find the argument that mining is wrong because the earth provides everything we need on the surface and close to hand. Here again is the other objection implied by Ovid, that mining is a profanation of the places where the spirits of the dead reside. But Pliny realises for us this horror at the invasion of the earth by suggesting that tremors and the fissures in the earth occasioned by earthquakes are an expression of the anger which the earth feels at being mined. This suggestion is touching enough as it stands, but it is made all the more poignant by the circumstances of Pliny's own death. He died in the great eruption of Vesuvius in 79AD which destroyed Pompeii. For Pliny, as for Ovid, the Rome of earlier times was untainted by the love of gold:

> The worst crime against man's life was committed by the person who first put gold on his fingers.

He pointed out that at Rome gold was hardly valued in the early period. In 390 BC, for example, no more than a thousand pounds weight of gold could be found in the city to buy peace from the Gauls. Not until much later still did the Romans use gold in their coinage, the second greatest crime in Pliny's opinion after the making of gold rings. With the invention of money came usury and a profitable life of idleness. But Pliny was powerfully aware of the attraction of gold and gives a brilliant account of its special properties among the metals. His account of the metals in general, their qualities, their therapeutic and other uses, and the means of mining them, is the finest in ancient literature.

Pliny and Ovid objected to the use of the metals and the means of winning them in the belief that the world was a better place before these technologies were discovered. For Ovid and other poets, that better time belonged to a far distant and perhaps imaginary past when Saturn ruled Latium and the earth provided everything needful without human effort. For Pliny it seems to have been rather a question of degree: the arts of mining and working metal were unnecessary, burdensome and harmful in comparison to the other means by which humankind provided for itself. But in Pliny, too, there is nostalgia for the Rome of earlier times. The attacks of these authors gain their force, therefore, from their belief in a better, even idyllic, past in which people lived simpler, more moral lives. Underpinning this belief was another, that a benevolent creation provided everything necessary for human life on the surface of the earth, that the essentials were easy to obtain. For Pliny and Ovid as for the Chinese sage

> Things hard to come by
> Serve to hinder our progress.

Ovid's and Pliny's strictures on mining and the metals have evoked some fascinating responses in more recent times, particularly in the sixteenth century. At this time they were considered in detail by two famous authors, Thomas More in *Utopia* and Georgius Agricola in his compendious work on the metals *De Re Metallica*. In his account of the life of the Utopians, More repeats some of Pliny's objection to mining when he tells us that the Utopians have no respect for silver or gold, believing that kind mother nature has deliberately placed all her greatest blessings close to hand, and hidden out of sight what is of no use to us. But for the Utopians iron is of value, is indeed as necessary to human life as fire and water, and in this respect More departs from the unqualified condemnation of the metals which we find in the earlier writers. The Utopians go to great lengths to ensure that gold and silver are as despised as they should be. They realise that if they locked them away, their fellow citizens would think they were being cheated. So instead they use gold and silver for the most base purposes, to make chamber pots for example, and the chains and fetters of slaves. Anyone who has committed a really shameful crime in Utopia is forced to wear gold rings in his ears and on his fingers, a gold necklace and a gold crown. Similarly with jewels. These they give to children as toys who soon grow out of them, and from then on the children despise them as the playthings of infants.

These customs of the Utopians have some odd consequences. On one occasion, for example, the Utopians received in their capital city an embassy from the Flatulentines. Flatulentia was some way from Utopia and the Flatulentines knew little of Utopia except that its citizens lived very simple lives and wore the same sort of clothes. So the Flatulentine embassy went out of its way to impress the Utopians by their splendour. There were only three diplomats in the party but they brought with them a hundred gorgeously dressed retainers while they themselves wore cloth of gold, gold chains and rings, and hats covered with jewels and pearls. All the Utopians in the city came out to see them, but because of their different customs the Utopians failed to recognise the diplomats and paid their respects instead to the humblest and least distinguished members of the embassy. They assumed that the diplomats themselves were slaves because of their gold chains.

> You might have seen children, who had themselves thrown away their pearls and gems, nudge their mothers when they saw the ambassadors' jewelled caps, and say:
> "Look at that big lummox, mother, who's still wearing pearls and jewels as if he were a little kid!"
> But the mother, in all seriousness, would answer:
> "Hush, my boy, I think he is one of the ambassador's fools."
> Others found fault with the golden chains as useless, because they were so flimsy any slave could break them, and so loose that he could easily shake them off and run away whenever he wanted.

More's objections to mining follow Pliny and Ovid insofar as all three authors suppose that delving into the bowels of the earth is contrary to the natural order, and that the precious metals incite people to crime. And like Ovid, More presents his case in the context of an ideal human society which is contrasted with the corrupt condition of his own. To a greater extent than any other human society, the Utopians live by the light of reason, which leads them to forms of conduct so much at odds with those of other societies that they startle the reader and appear at first absurd. More enjoys these paradoxes to the full. All through his account of Utopia there is an element of play as he generates the most extraordinary forms of human behaviour from principles which appear quite reasonable and straightforward. Georgius Agricola, on the other hand, approaches the ancient criticisms of mining and metallurgy as a practitioner who is seriously worried and annoyed by them. There are no funny stories in his account. He sets out the case against mining and metallurgy in great detail and then proceeds to answer it point by point. He puts the arts of mining and metallurgy on trial and gives, in full, the cases for and against. This first book of his *De Re Metallica* is a masterpiece of rhetoric in which we find many of the arguments which are used today.

Agricola quotes the lines from Ovid's *Metamorphoses* about mining, and then answers Ovid's objections by claiming that those who speak ill of the metals and refuse to make use of them are in fact abusing God who created them. For if the metals are useless, then God erred in creating them. But to believe this is impious. It may be that Agricola's argument here would carry more weight with Jews and Christians than it would in the classical world, since in Jewish and Christian scriptures God's creation of gold in the garden of Eden is explicit. As for the argument that mining contravenes the natural order, Agricola replied that the metals are not hidden in the earth to keep us from them; they are where they are because they are heavy. Even if they were generated in some other element they would still sink to the earth. Of their very nature they are formed within the depths of the earth, and the mining of them is much more natural than, say, the catching of fish, since the earth is the proper element of the human race. He considers the argument that mining devastates the land and points out that most mines are on the sides of mountains or in gloomy valleys which are of little use for agriculture. Certainly the mining of such places disrupts the plants and wildlife, but the profits from mining can be used to restock these sites when the mining is complete.

There are, however, two objections to mining which Agricola raises but does not quite answer, and both of these objections have a distinctly modern ring. When the ores are washed, the detractors of mining argue, the water which has been used poisons the streams and rivers and either kills the fish or drives them away. This is an early reference to the problem of pollution and it is an argument against mining which Agricola overlooks when he comes to his defence of it. On the other hand his general counter argument against those who claim that mining devastates the environment, that the profits from mining may be used to restore such damage when the mining is completed, may be taken to apply to this special case of water pollution as well. He also raises the objection that mining, unlike agriculture, is an unstable profession by its very nature since the mine, unlike the field, is soon exhausted. Here we have the beginnings of the argument that mining is wrong because it exploits resources which do not renew themselves. Pliny also deplores this in the passage quoted. Agricola gives a partial answer to this by pointing out that many mines have been worked for several centuries and are still producing valuable ore. He claims that while mines are eventually exhausted, they are nonetheless so much more productive of wealth than agriculture when they are in use, that this way of using the land is justified.

Agricola is most effective when he deals with the claims that the metals are evil because they are used in war and because they incite people to crime. He answers these objections in the same way. In neither case, he argues, are the metals to blame. He demonstrates at length how the use of the metals in war is hardly the fault of the metals but of those who use them so. Such people are quite capable, he suggests, of the most terrible cruelty without the metals, and he lists a horrifying series of tortures in which the metals play no part. As for the claim of Ovid and Pliny that the precious metals incite people to crime, he argues that this is no more the fault of these metals than a woman's beauty is at fault for whom a man commits a crime. It is not the woman's beauty, but the man's unbridled lust which is to blame. On these grounds he dismisses as unintelligent all the authorities and poets cited by those who object to the precious metals, and then produces his own authorities from whom he quotes many passages which praise wealth when it is wisely used.

Agricola's presentation of the case against mining comes as a surprise. We are in the habit of thinking that the resistance to mining is a recent phenomenon, but here from the sixteenth century we are presented with one of the most complex attacks on mining to be found in the literature. There is some evidence that this negative way of regarding mining was much more widespread in Europe than this cursory summary of some poets and writers would suggest. In certain periods of Roman history mining was banned in Italy because of its harmful effect upon agriculture, and in more recent times the great mines of Spain were closed after the discovery of the mineral wealth in the new world. These societies, it seems, were very ready to prevent mining on their own soil as soon as other mining sites became available to them. It appears, therefore, that debates over the morality of mining have persisted for many centuries in the west, and that Agricola is already the heir to a long tradition. From this point of view the contemporary attacks on mining are not the result of a sudden realisation by a much more enlightened society; it is one more skirmish in a battle which has been going on for millennia.

The arguments in the *De Re Metallica* sound contemporary in another way too. Though he quotes Ovid's lines which criticise mining as a delving into the bowels of the earth, Agricola does not answer them at the level at which they are intended. For Ovid and Pliny there is something dreadful about this invasion of nature. It is profoundly wrong in itself, regardless of its consequences. But Agricola considers the consequences only. His argument that mining is usually conducted on mountains or in gloomy valleys which are of no use to us otherwise, would not weigh very much with Ovid or Pliny. For them the act of mining is a desecration of nature wherever it is conducted. The earth is not there simply for our use, but has a certain spiritual or numinous identity which is seriously damaged by mining. Agricola's handling of this objection betrays a characteristically modern incapacity to comprehend it. This is the more surprising in an author who was deeply interested in the fairy denizens of the mine and wrote a book about them. It may be that his lack of sympathy with the objections of the classical authors stems from his Christian background. For in the Judaeo Christian scriptures, the earth is indeed placed at the service of humankind to do with it as we will. That the creator made the metals and put them in the earth is for Agricola a sufficient justification for our retrieving them.

The sense of the divine spirit in nature as we feel it in the writings of Ovid and Pliny, and the incapacity of the later Christian imagination to comprehend this, continue to inform and deform the debate in recent times. When we take into account the very limited area of land affected by mining directly and the temporary nature of the damage done to it, it is otherwise inexplicable that the act of mining should arouse the fears it does. The explanation for these fears must be that they spring from an intuitive horror at the destruction of life on the earth's surface and the breaking through of the topsoil. Mining is considered a kind of rape, a violation of the mother, and this perception is reinforced in our time by the image of massive bulldozers and scoops in the process of excavation. For us it is extremely difficult to articulate this vision since the language in which to theologise nature has largely been lost. Wordsworth, who spent his poetic life struggling to reclaim it, described this difficulty in some memorable lines from his sonnet on the corruption of the world by business:

> Great God! I'd rather be
> A pagan suckled in a creed outworn;
> So might I, standing on this pleasant lea,
> Have glimpses that would make me less forlorn;
> Have sight of Proteus rising from the sea;
> Or hear old Triton blow his wreathed horn.

Wordsworth here compromised his hope of salvation in Christ for the chance to see his natural surroundings as the Greeks had seen theirs. The scale of the risk he was prepared to take measures exactly the intensity of his despair.

This same despair dogs present debates over the damage done to the environment by mining. With the loss of the old pagan spirits of nature and the departure of the fairies, the problem of how to articulate the horror people feel for mining remains acute. Some environmentalists have made a half hearted attempt to reconstruct the old sense of the spiritual in nature in terms of a global spirit of the earth. This is usually called Gaia, the Greek name for the goddess of the earth. But since the old myths and rituals have long since disappeared, this has not been successful. Instead, these environmentalists tend to change their ground at this point in the argument, lapsing quickly from considerations of how nature should be revered in itself to calculations of self interest. The ecosystem, they claim, should be left untouched because damage to part of it will damage the whole, and thereby threaten us. Above all, their conception of Gaia lacks any real sense of place, since the increasing mobility of the modern world makes it very difficult for more and more of us to feel of any one place that this is our home, our mother and nurse.

In Australia, however, a surrogate has been found which enables us to articulate our sense of the earth's inviolability and sacredness. This surrogate is the Aboriginal sense of the spirits of place. In the confrontation between mining interests and those who argue for the preservation of the Aborigines' sacred sites, the ancient horror of Ovid and Pliny is realised anew. In the mythology of the Aborigines and their song cycles, the geographical features of the land play a large part. The destruction of a mountain by mining cannot in this context be regarded as the temporary disruption of an otherwise useless wilderness. It becomes instead the irreversible removal of a living symbol in which the identity of a whole tribe has been located for many generations. In this way the landscape is invested with a primordial power which has accrued to it over many centuries, and this age old power is set against the short term wealth to be gained from its transformation by mining. In such ways the early inhabitants of the continent cast a kind of shadow across the consciousness of those who have come after, recalling us to older and deeper feelings for the land. Many respond to this call because they too feel that there is a deeper significance in the land than our reason allows. Though there is much in the traditional patterns of Aboriginal life which we are happy to see discarded or reformed, at this point we stop and wonder whether here they would do well to reform us. Our concern here is not primarily to protect their rights and customs, but to acquire their understanding of the land for ourselves.

This task is probably beyond the rest of us. However much we may wish to borrow this perception of the land from its original inhabitants, our capacity to do so is very limited. Such a perception may answer to certain needs of ours, but we have none of the customs, rituals, children's games, tribal memories by which it is instilled and reinforced in those with whom we would share it. We remain on the outside, looking in. To see the land as an Aborigine sees it requires of us so radical an alteration in the understanding that we can hardly begin to imagine it. At the very least the Aboriginal view of the land presupposes a metaphysics in which all the phenomena of experience derive from another realm, a different order of reality. This 'dreamtime', as we call it, has certain parallels with elements in western thought, but these are just the elements which are most heavily discounted in our time. It is hard enough for us to comprehend the Platonic theory of ideas or the Christian Logos through whom the world was made, despite the fact that these are central to the intellectual history of our own culture. The task of understanding how the Australian landscape manifests such a metaphysics is very much more difficult again.

For these reasons the Aboriginal understanding of the land must remain for us a largely unrealisable ideal. The existence of these people among us is a constant reminder of other ways of thinking and feeling, over which our incomprehension casts a special glamour. Just as Ovid, Pliny and Virgil looked back to a mythical time in which Saturn ruled in Italy and everything that people needed was provided with no effort on their part, just as More imagined a Utopia in which everyone lives more simply and more rationally than anywhere else, so the world of the Aborigines has come to represent a standard in contemporary Australia against which our own ways of doing things are felt to fall short. But More's Utopians valued iron as essential to human life despite their belief that mother nature had hidden deep in the earth what we do not need. Similarly the Aborigines of Australia traditionally valued certain minerals and rocks which they mined and quarried with enthusiasm, the ochres which they used to paint their bodies and the stones for their weapons and tools.

> Cosmetics gave a wonderful amount of pleasure. The very thought of them seemed to excite aboriginals. The Englishman, George Robinson, was taken to the mine near Mount Roland by aboriginals in the winter of 1834 and he observed how enthusiastically they began to mine the rock as soon as they arrived, 'being quite overjoyed at the sight' of the quarry. The women were usually the miners, at least in the twilight of Tasmanian life, and they levered out the iron ore by gripping a stone in their hand and using it to strike a sharp pointed stick into the rock. Robinson was amazed to see the signs of strenuous working - the old excavations, the heaps of stone discarded on the slopes, and narrow holes with collapsed sides. He watched the way in which women squeezed themselves down narrow excavations - one became stuck in the crevice and had to be pulled out by the legs. When the ochre had been packed in the kangaroo skins kept for that purpose, the women set out

slowly, some carrying large loads which would have tested the strength of a man. With the ocean visible below them, and to the right the blue hills standing out across the Tamar, the women walked slowly along a welltrodden track with enough iron ore on their backs to furnish cosmetics - if used frugally - for ten thousand faces. As always they squandered the red ochre with the extravagance of millionaires.

Though the Aborigines did not smelt the ore bodies so as to extract the metals, there is no reason to suppose that they abstained from doing so. Their understanding of the world around them need not have prevented their incorporating techniques of metallurgy into their patterns of ritual and belief. We must be careful not to attribute to these people our Wordsworthian view of nature as untouchable. The Aborigines' closeness to the land was a matter of use as well as ritual, and they were very ready to destroy the surface growth and the creatures which depended on it by setting fires.

It is a mistake therefore to suppose that the mining and smelting of metals is a violation of nature according to the Aboriginal ethos. This is not where the line should be drawn in the confrontation between mining and sacred sites. As we shall see, the bringing to maturation of the metals in the ground through mining and smelting was traditionally regarded in the west, both in pagan and Christian times, as a fulfilment and consecration of nature and not as its defilement. This way of thinking is ignored in the authors we have studied in this chapter, not only by Pliny and Ovid but even by St Thomas More and Agricola. The problem over the Aboriginal sacred sites arises as it does because this traditional way of regarding work in the west has been lost; the sacred cooperation between the human and natural worlds through craft is now almost entirely forgotten. This desacralisation leaves us with no other way of thinking about mining than as a violent seizure of wealth from the earth. The juxtaposition of mining, when considered in this way, with the Aboriginal rituals of the earth puts us in the wrong.

How one people developed a ritual of metallurgy is set out in the following passage from a Senegalese autobiography. This passage exemplifies how the working of gold, at least, has continued to be regarded as a sacred art right up to the present day among some African peoples. This suggests that the confrontation between miners and the defenders of sacred sites does not originate in the desecration of those sites so much as from the desacralisation of mining and metallurgy. Where these practices have remained sacred themselves, the conflict over this use of the land cannot become the problem it has for us.

On a sign from my father the two apprentices started working the sheep-skin bellows, which were situated on either side of the forge and connected to it by means of clay pipes... The flames in the forge shot up and seemed to come to life, an animated and evil genius.

My father then grasped the smelting-pot with his long tongs, and placed it on the flames.

All of a sudden all other work in the smithy was stilled, for during the time that gold is being smelted, and while it cools, it is forbidden to work either copper or aluminium in its proximity, in case even a particle of these base metals should enter the smelting pot. Only steel may continue to be worked. But even those engaged on a task with steel would usually finish it quickly or lay it aside, in order to join the apprentices gathered round the forge.

When my father felt that his movements were being impeded by the apprentices crowding round, he would silently motion them to stand back. Neither he nor anyone else would utter a word. No one dared speak, and even the minstrel was silent. The stillness was broken only by the wheezing of the bellows and the low hissing of the gold. But though my father said not a word, I knew that he spoke inwardly; I could see that from his lips which moved silently as he stirred the gold and the charcoal with a stick, which, as it caught fire, he had to keep replacing.

What could he be saying inwardly? I cannot say for sure, as he never told me. Yet what could it be but an invocation? Did he not invoke the spirit of the fire and of the gold, of the fire and of the wind, of the wind which blew through the bellows, of the fire that was born of the wind, and of the gold that was wedded to the fire? Assuredly he summoned their help and entreated their friendship and communion; assuredly he invoked these spirits which are amongst the most important, and whose aid indeed is necessary for smelting.

The process which took place before my eyes was only outwardly the smelting of gold. It was something more besides; a magical process which the spirits could favour or hinder. That is why stillness reigned around my father.

Was it not remarkable that at such a moment the little black snake always lay hidden under the sheepskin? For it was not always there. It did not come and visit my father every day, yet it never failed to appear when gold was being worked. This did not really surprise me. Ever since, one evening, my father told me of the spirit of our tribe, I had found it quite natural that the snake should be there, for the snake knew the future.

SLAVE MINING AND THE RIGHT TO WORK

Another long standing attack on mining which Agricola considers in his *De Re Metallica* is that mining dehumanises miners. On this view mining is not an activity for free men but for slaves. Agricola's answer to this is that agriculture too is sometimes the work of slaves but that does not make it unworthy of the free. On the contrary, Agricola argues, of all the crafts and professions, mining has some claim to being the most honourable, since it is more profitable than farming, more honest than the work of the merchant or moneylender, less violent than soldiering and so on. We can hear in this echoes of the medieval strictures on those means of acquiring wealth which do not copy the divine creativity of nature. It is certainly the case that in Agricola's Germany and many other places, the profession of the miner has been highly regarded, but it is also true that mines have very often been the scene of the most brutal human exploitation. As a result the image of mining as a human activity has oscillated between two extremes, and in this respect mining is an exception among the various forms of work. The only other craft or profession with a similarly ambiguous status is that of the soldier.

The working life of the miner was rarely described by ancient authors except where conditions were so egregiously bad that they aroused the author's sympathy for those subjected to them. Such was the case with the gold mines in southern Egypt and the silver mines in Spain as they were described by Diodorus Siculus. Writing two centuries before the Christian era, Diodorus gives a very detailed account of the Egyptian gold mines and of those who worked them. He tells us that these mines were very ancient but for how long the working conditions which he describes had persisted he does not say. He explains that the kings of Egypt reached an agreement among themselves to condemn criminals and captives of war to these mines, as well as many others who were unjustly condemned and occasionally all their families as well. In this way these kings acquired enormous revenues at very little cost. All those so condemned wore chains and were made to work day and night without ceasing, the strongest of them underground and in darkness except for the lamps on their foreheads. Diodorus specifies that their work was entirely unskilled, requiring only brute force as they cut tunnels through the stone wherever the veins of gold led. These operations were carried out at the direction of a skilled worker who told the labourers where to dig, while an overseer compelled them to unceasing efforts with blows.

Outside the mines the older men, boys and women processed the ore by grinding it to the consistency of the finest flour. No opportunity was given these people to exercise the slightest care for their bodies nor were they allowed any clothes. No one, Diodorus writes, could look upon these unfortunate wretches without feeling pity for them because of the terrible hardships they suffered. No leniency or respite of any kind was given to the sick or the old or women, but all without exception were forced by blows to persevere in their labours until they died. In the final stages of the process skilled workers received the powdered stone and by various means separated whatever was porous and earthy until only the gold dust remained. Diodorus concludes his account of the Egyptian mines with the observation that the production of gold is laborious, the guarding of it difficult, the desire for it enormous and its use half way between pleasure and pain.

Diodorus' account of the Egyptian gold mines attracted the attention of Karl Marx who understood how rare such working conditions were in antiquity. Nonetheless Diodorus gives a very similar description of the working conditions in the Spanish mines under Roman administration. The earliest period to which we can date this way of using slaves in mines is the fifth century BC. At this time the Athenians began to exploit the very rich deposits of silver at Laureion in Attica. It appears that these deposits had been discovered very much earlier but had been largely left unworked because free men refused to work underground, and because the capital investment required to sink shafts was not worth the risk of frequent failure. But throughout the fifth century, at the very same time as Athens was becoming more and more democratic, a new way of exploiting the mines at Laureion was introduced. We must remember that throughout the fifth century Athens was under almost continuous threat, first from the Persians and then from Sparta and her allies during the Peloponesian war. The Athenians' need for money, to defend Attica against invasion and to prosecute their expansionist policies, drove them to employ a new class of slaves in their mines. These were men who were of no use to their owners except for their physical strength. Their owners hired mining concessions from the state, then put their slaves to work in the mines under skilled overseers to whom they paid high wages or for whom they paid high prices as slaves.

The historian Alfred Zimmern sets out these new conditions of work in the following terms:

> Unskilled underground mining is in fact a class of work which lends itself most conveniently to the perfect form of chattel slavery. All that is required of the slave is a vigorous body, and sufficient of that lower kind of reason which, Aristotle tells us, is necessary to understand a spoken command; all that is required of the master is watchful and drastic supervision, or sufficient capital to provide efficient overseers to do this for him. The work is mechanical, unchanging, practically inexhaustible, and entirely unskilled. The workers are almost stationary in their places and can be chained without interfering with their efficiency. They work with only the roughest tools and appliances. The work does not involve disease (which would mean loss of capital), but is yet sufficiently exhausting to lower the vitality and so make it likely that death will follow closely upon the failure of working power. It is

carried on in a number of separate pits and galleries underground, under conditions where the amount of work performed can easily be measured and tested, and where the task of supervision is extraordinarily simple and inexpensive. The overseer (generally a trusted superior slave) could probably look after the entire property of a considerable mine owner or concessionaire. Above all, it is expended in production of silver, almost the only article for which there can be said to have been an international market and an unlimited demand.

According to Zimmern, all this represented a massive break with older Athenian traditions of work. The silver so produced was coined and stamped with the Athenian owl as the miners were branded with their masters' marks. The uniformity of the coins corresponded exactly to the loss of any personal character in the work of the miners. Everything else produced by the industry of slaves had an individuality of its own, but the owls of Laureion testified only to a new kind of industrialism in which the personality of the individual worker was reduced to nothing. And this was occurring just as the first democratic state was being organised in the very same place. We may perhaps compare this to the fine fury of Wilberforce's anti-slavery movement at the beginning of the nineteenth century at the same time as the working conditions of English men, women and children reached their nadir in the early phases of the industrial revolution.

It is very hard to assess how far these working conditions were typical of mining in antiquity and later. Certainly we know of other civilizations in which matters were arranged very differently. In Peru, for example, just before the Spanish conquest, the working conditions of miners were very carefully regulated by the state which kept a register of the population in the mining provinces and of the mineral resources. The Incas ensured that while the most competent people were selected to undertake the mining, the weight of the work should not fall disproportionately on any of them. The miner was provided by the state with all that he needed for the work and the length of his service in the mine was exactly stipulated. By a constant rotation of labour, no one was overburdened and each had time to provide for the demands of his own fields and household. In this way the health of the miners was preserved even when the work was most wearing and unwholesome. All this stands in the sharpest contrast with the conditions under which these same mines were worked by the same people under Spanish administration. For then the appalling conditions of the slave mines in antiquity were reproduced and a large part of the indigenous population was systematically worked to death.

In Europe, meanwhile, conditions for miners were often better than they were for other workers. In England, France and Germany, miners were given special privileges from medieval times. In 1201 English miners were protected from outside interference by royal decree. They formed strong organisations whose legal and judicial acts could only be reversed by the warden appointed by the king. At the end of the thirteenth century the French king freed the serfs and ended all forced labour in French mines. In the sixteenth century in Saxony, the working day of the miners and smelters was standardised, and the double shifts which some miners had worked in the middle ages were prohibited. At the same time provision was made for the payment of wages to miners up to two months off work because of injury. Miners' associations cared for the poor, disabled and retired miners and organised their funerals, in which the youngest brethren carried the coffin. These coffins were decorated with the emblems of the miners' work, as well as with shrouds and palls belonging to the miners' association. The earliest of these associations known to us is the Society of Hewers in the Saxon capital of Freiberg from 1400 onwards. The social status of the miners and metallurgical workers was generally as high as that of the citizens in the rising towns, according to the *Cambridge Economic History of Europe.* Very little social distinction seems to have been made between the miners and those who provided the capital. The rulers of these mining areas took great care to foster the development of a skilled and self perpetuating work force.

In Australia, likewise, miners have struggled for and maintained a fierce independence. On the goldfields of Victoria they fought against the encroachment on their rights by the state and created an effective political organisation. In the Red Ribbon riots in Bendigo and in the Eureka stockade at Ballarat, where many miners were shot dead by soldiers, they showed a determination and courage which few other crafts or professions have equalled. The unavoidable hardships of their labour underground, the risks which they daily encountered, made the diggers tough. Miners have always been regarded as particularly suited for soldiering when they could be spared from their proper work; and in the first world war Australian miners gave their name to the entire Australian army. From then on all Australian soldiers were known as diggers. In these ways the working life of the miner came to be a paradigm of manly, free and independent action, a conception of their work which stands at the opposite extreme to that of the slave mines. With the development of technology, mining has come to require more and more capital and less and less human labour. The loss of that labour force, the disappearance of most of the miners, may have been unavoidable, but its political consequences have been serious. How differently would the recent debates about mining have gone in Australia if, like timber workers, miners had been a strong and independent labour force anxious to defend their right to work?

These instances show that the miners' labour, though hard and dangerous, is no more degrading than any other forms of work. For Agricola mining was a most profitable profession and a most honourable one, and he encouraged those who invested in mines to take part in the actual working of them. Even in those mines where conditions were worst, as in the Spanish silver mines, writers on mining found much to admire in the ingenuity with which the deepest mines were drained. As far back as the book of Job in the Old Testament, the skills of miners were chosen as the supreme example of human wit in action, as contrasted with the wisdom of God. For Job, it seems, mining best demonstrated the ingenuity which distinguishes the human from other creatures:

> Surely there is a mine for silver,
> And a place where gold is refined.
> Iron is taken from the earth,
> And copper is smelted from ore.
> Man puts an end to darkness,
> And searches every recess
> For ore in the darkness and the shadow of death.
> He breaks open a shaft away from people;
> In places forgotten by feet
> They hang far away from men;
> They swing to and fro.
> As for the earth, from it comes bread,
> But underneath it is turned up as by fire;
> Its stones are the source of sapphires,
> And it contains gold dust.
> That path no bird knows,
> Nor has the falcon's eye seen it.
> The proud lions have not trodden it,
> Nor has the fierce lion passed over it.
> He puts his hand on the flint;
> He overturns the mountains at the roots.
> He cuts out channels in the rocks.
> And his eye sees every precious thing.
> He dams up the streams from trickling;
> What is hidden he brings forth to light.

Diodorus in his account of the Egyptian gold mines states explicitly that the work enforced from those underground required no skill whatever. This point is taken further by Zimmern in his description of the conditions in Laureion. Mining is a very skilled profession and even where much or most of the labour is carried out by unskilled workers, the knowledge required to direct such work should be valued. The best working conditions are those which develop this knowledge and these skills to the highest degree in the greatest number of people. Knowledge as knowledge is a good in itself; those who have it are the better for it. The knowledge acquired by the prospector, the geologist, the miner, the assayer, the metallurgist is of the greatest benefit to these people. But, insofar as their knowledge is of the earth and its constituents, this knowledge directly benefits the earth also, since it brings the human and natural worlds into close union and realises certain potentialities in the natural order.

These aspects of work in general and of mining in particular were ignored by the early critics as they are ignored by more recent ones. And this despite the fact that these aspects were presented often and with great clarity from the earliest periods of both the Greek and the Hebrew traditions. But they do not rate a mention by Ovid or Pliny or in More's Utopia. The rest of this book is another attempt to establish these traditional values of our working with metals. I will show how the metal crafts have been sanctified and made ideal from earliest times. And I will argue that to criticise mining in most of the ways which have now become common is not so much a means to our recovery of the spirit as it is an active rejection of the spirit.

Chapter Two

THE CLASSICAL TRADITION

A kind of radiance, like that of the sun or moon, lit up the high roofed halls of the great king. Walls of bronze, topped with blue enamel tiles, ran round to left and right from the threshold to the back of the court. The interior of the well built mansion was guarded by golden doors hung on posts of silver which rose from the bronze threshold. The lintel they supported was of silver too, and the door handle of gold. On either side stood gold and silver dogs, which Hephaestus had made with consummate skill, to keep watch over the palace of great hearted Alcinous and serve him as immortal sentries never doomed to age. Inside the hall, high chairs were ranged along the walls on either side, right round from the threshold to the chamber at the back, and each was draped with a delicately woven cover that the women had worked. Here the Phaeacian chieftains sat and enjoyed the food and wine which were always forthcoming, while youths of gold, fixed on stout pedestals, held flaming torches in their hands to light the banqueters in the hall by night.

Homer

O F ALL THE ANCIENT civilisations we study now, the Greek was the most committed to the ideals of beauty and craft. For the ancient Greeks the pursuit of beauty seems often to have been a matter of more concern than truth or wealth or even life itself. Their notion of what was beautiful extended well beyond our own. Beauty was not just a property of the physical world, of people, places and things. It was a property of souls, of laws, of institutions, of whole sciences, and especially of their gods and goddesses, whom they represented as physically beautiful, but whose real beauty was of the spirit.

The greatest hero of the Greeks was Achilles, and part of his story is told in one of the first and greatest of all Greek poems, the *Iliad* of Homer. Physically Achilles was the most beautiful of all the heroes who fought at Troy, in the war which was fought for Helen, the most beautiful woman in the world. But the real quality of Achilles shone out in his actions. He was given the choice of a short and glorious life or a long and inglorious one, and he chose glory. By doing so he established a standard of beauty or, as we might say, of nobility and fineness, which inspired the whole of Greek civilisation.

The Greeks' cultivation of beauty can be seen clearly in the ruins of their temples and in the remains of their handcrafts which we have preserved in our museums. But their love of craft is clearer still in their literary and philosophical remains. Greek philosophers developed the most thorough and far reaching theory of craft to be found in western thought. This theory emphasised the relations between human art and the nature and power of the divine. It explained the common Greek practice of attributing to gods and goddesses the patronage of the various arts and crafts. According to this theory, the art in the artist or craftworker is far greater and more beautiful than the art in the thing done or made. From the beauty of what is done or made we catch a glimmer of the beauty which the artist conceived, contemplating the idea of the work in the mind of God. For the Greek the means by which a work of art came into existence was always miraculous, to be explained only by supposing a direct inspiration from the divine. The artist or craftworker was a medium through whom the divine powers of creation acted upon the world.

This is why the poet Homer invokes the Muse at the beginning of his two long poems, the *Iliad* and the *Odyssey*. It is not he who composes these poems but the goddess. He is merely the medium of their transmission. Conscious as he is of the source of creation within himself, he is also the greatest poet of the arts and crafts in the history of literature. There is hardly a page of the *Iliad* or the *Odyssey* where we are not brought face to face with the miracle of creation, or with the efforts and aspirations of an artist. An honest working woman weighs the wool she has carded in her balance with agonising care, anxious not to give short weight, nor to give the tiniest amount more than she needs to for the meagre pittance with which to feed her family. Just so does a battle between the Greeks and the Trojans hang in the balance. A red hot wooden stake thrust into a giant's eye hisses like iron dropped from the forge into water, to temper it and give it strength. At a tense moment the battle front between the Greeks and Trojans sways no more than does the line that is stretched along a ship's timber to test its truth by a skilful carpenter who has mastered his trade in the school of Athene.

But metaphors and similes are comparatively rare, even in the poetry of Homer who is famous for them. More striking still is the way in which everything in Homer's world is well made or well done, glittering and resonant at the very limit of its potential. Never has there been a poet more conscious of the skill and care which have gone into the making of even the least of the works of craft. The furniture is better made, the clothes more beautiful, the speeches more grandiloquent, the weapons more splendid than anything we know of here. To hear him describe how a king dresses himself or how a meal is prepared is to wonder at the art of these things as if for the first time. Despite the savagery and slaughter of the battle scenes, it always seems as if the world of Homer were but a single step from paradise. The only question is whether the world he describes is really better than ours, or merely seems so because of his greater sensitivity to it. Certainly the reading of Homer helps to restore that joy in our surroundings which is our proper state of mind.

Among all the works of craft with which Homer deals, the metals are central. For Homer the art of the smith is the prime example of craft work. Sensitive to everything else that is done or made, Homer is exceptionally sensitive to the working of metal. He is preoccupied with whatever glitters, sparkles, dazzles. His vocabulary for the shining of things is enormous. Sometimes the effect is frightening as when a little boy is terrified by the bronze of his father's helmet. Sometimes it is mesmeric as when Achilles appears in the armour just made for him by the god Hephaestus. At the very instant when Achilles strikes down his greatest enemy, the spearpoint with which he does it shines

> like the Evening Star among the stars at the milking time of
>   night,
> the most beautiful star to stand in the sky.

Most often of all, the effect of the metals in Homer's poetry is straightforwardly magnificent, as it is in the passage from the Odyssey with which this chapter begins.

In this description of the palace the effect is precisely calculated as silver is added to bronze and gold to silver until the passage culminates in the golden youths with their flaming torches lighting the feasters through the night. The effect achieved is one of infinite reduplication. The golden youths replicate each other and each of the torches which they hold is reflected from the youth who holds it, from the other youths and from the bronze and gold and silver which otherwise surround them. A radiance like that of sun or moon comes from the palace, especially at night when the palace is lit entirely from within. We note that the metals used in the construction of the doorway are each of them used in the most practical way. There is nothing in the archaeological record to suggest that Homer could have known a palace like this in his own time, if he lived in the ninth or eighth centuries BC as is generally supposed. Perhaps he inherited this description, as he inherited other things, from the generations of ministrels who had sung before him, so that we have here the record of a palace from

Mycenean or Minoan times, though much elaborated. Still other possibilities arise from the ancient tradition that Homer was blind. Is he preoccupied with the metals because they are the brightest and sharpest sights from a visible world he can now only remember? Could that palace, shining by night, serve as a metaphor for the inward imagining of the poet himself, the more brilliant because no longer obscured by physical sight?

HOMER'S SMITH GOD

There are the metaphors from the crafts and there are the descriptions of craft objects. But Homer's preeminence as a poet of the crafts and of metallurgy is also to be seen in his descriptions of the god Hephaestus. Hephaestus made the gold and silver hounds which guarded the palace of King Alcinous. As the god of smiths and the most prominent craftworker among Homer's Olympians, Hephaestus is the paragon of all who work with metals. He is also the most amicable and endearing of all the gods and goddesses on Olympus. His character and his presence on Olympus are clear evidence that Homer rated metallurgy above the other crafts. Since Homer was the most influential teacher of the Greeks, we may take it that this was true of ancient Greek civilisation in general. The Romans, too, inherited Hephaestus from Homer, renamed him Vulcan and developed him further.

The fullest account of Hephaestus is in the eighteenth book of the *Iliad.* At this point in the story the hero Achilles has lost his armour, yet is eager to fight. So he prays on the seashore to his mother, the sea nymph Thetis. Thetis hears his prayer and rises out of the sea to help him. She promises to go to Olympus, as she can since she is a nymph, and there ask Hephaestus to make him new armour. Thetis then flies to Olympus. She arrives at Hephaestus' palace which shines like a star and is made of bronze, the most beautiful of all the palaces of the gods, as is fitting for the god who made them all. There Thetis is warmly greeted by Charis, Hephaestus' wife. Thetis is clearly a great favourite in this household and is given an overwhelming reception. Charis sits her down in a place of honour and reproaches her for not coming more often. Then Charis calls to Hephaestus where he is hard at work in his workshop, and tells him who has arrived.

'That great and wonderful goddess here!' Hephaestus exclaims and straightaway gives a long description of how Thetis once saved his life. At this point he is still in his workshop and must be speaking to the two goddesses from a distance. He tells how he was cast out of heaven because of his lameness by his mother Hera who wished to hide him. He was in agony as a result of his fall and in danger of his life, but Thetis had taken him in and hidden him and looked after him, and for nine years he had stayed with her in her hollow cave. For nine years by the stream of murmuring Ocean he had forged wonderful things in her cave

bronze ornaments, brooches and spiral armbands, rosettes and neck-
laces

and nobody had known he was there. And now, he declares, he is very
anxious to pay Thetis the full price for having saved his life. We may take
it that Thetis was well aware of his gratitude and was counting on it when
she promised her son Achilles that she would have armour made for him by
Hephaestus himself.

It is delightful to come across this scene between Hephaestus, Thetis
and Charis in the middle of a war epic. It is a relief to escape for a moment
from the horrors of the battlefield to the gentleness of Olympus. The hos-
pitality, the courtly manners, the sense of obligation, the absolute trust
between these three is the more inspiring after the bloody scenes which
have brought us here. It is sad to hear how Hera rejected her own child but
this is made up for by the kindness of Thetis. Hephaestus still feels himself
to be in her debt even though he spent the nine years in her cave making
beautiful trinkets for her. So unforced are the relations between these three,
so natural the setting in which their encounter takes place, that we could
easily forget that they are gods on Olympus and take them as simply human.

But Olympus is not in this world, not even at the top of a high mountain. It is a place beyond all the chances of weather, with a white radiance
playing over all. And the Olympians are immortals, beings unlimited by
the span of mortal life or mortal knowledge. Though Homer presents them
in a lifelike manner, their life is not ours. They are described in the language of this world because this is the only language we and the poet have,
but this should not mislead us into thinking that they are in all respects as
we are. At a certain point in the human story of Achilles, Thetis goes to
Hephaestus who tells her the story of their earlier companionship. At the
human level the events which he narrates are just passing moments. But in
the lives of gods such things are not incidental but essential; they reveal the
innermost nature of eternal powers. In this case the clue is given by the
catalogue of all the things Hephaestus made under the sea:

bronze ornaments, brooches and spiral armbands, rosettes and neck-
laces

The spirals suggest sea shells. When Hephaestus describes the wonderful things he made in Thetis' cave, he speaks as the great forger of all
the forms of marine life, of the spectacular world of nature beneath the
waves. These are the things he forged in his nine years' stay with Thetis,
and from our point of view he is forging them still. Hephaestus is the
divine creative power according to the Greek theology, who made every-

thing from the starry heavens to the forms of marine life. Everything we are told about his creations has significance well beyond its immediate context.

When Hephaestus has finished his story of how Thetis saved him, he gets up from his anvil, takes the bellows away from the fire and carefully puts all the tools he was using in their silver chest. Then he wipes his face and hands with a sponge and puts on a tunic to make himself presentable to Thetis. Homer gives us some details of the work he was engaged in when Thetis arrived:

> He was making a set of twenty cauldrons, each with three legs, to stand round the walls of his well built hall. He had fitted golden wheels to all their legs so that they could run by themselves to a meeting of the gods and amaze the company by running home again. They were not quite finished. He had still to put on the ornamental handles, and was fitting these and cutting the rivets.

As with the account of Hephaestus' work under the sea, his making of the cauldrons is presented as quite incidental to the main story, some details on the side to fill out the picture of Hephaestus as smith. And as with the account of his marine labours, so here Homer is giving us a far reaching theory of the creation.

Ancient commentators thought they knew very well what Homer meant by the twenty threelegged cauldrons which Hephaestus was making as Thetis arrived. They took it that Hephaestus was making a shape like one of those balls made of cloth or leather which are still made today. The ball is made by cutting out twelve pieces of material, all twelve pieces being of the same size and shape. Each piece has five equal sides. Stitch the twelve pieces together to produce the following shape which becomes almost perfectly round when it is filled or inflated. Of course, in the figure we see only half of the ball. There are another six pieces of the same size and shape hidden from us.

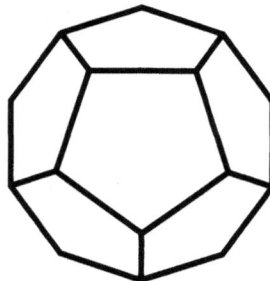

What has this to do with Hephaestus' twenty threelegged cauldrons? Each piece of our ball has five sides and each of these sides is joined to the side of another piece. Where these sides are joined, an edge is formed. Each line in the figure represents one of these edges. These edges also meet, and where they meet they form points or solid angles. Each of these points or angles is formed by the meeting of three edges, and on the surface of the whole ball there are just twenty such points. These twenty points at each of which three edges meet were taken to be the twenty threelegged cauldrons.

This way of interpreting Hephaestus' cauldrons, however unreasonable it may seem at first, has profound consequences. The figure in our drawing has one very special property: all its sides, edges and angles are equal. It shares this property with just four other solid figures of which the best known is the cube. These five regular solid figures were regarded by later Greek philosophers as the building blocks of the universe, the basic figures out of which everything else was made. The cube, for example, was taken to be the molecular structure of the element earth. The figure in our drawing was regarded as the fifth of the five regular solids and was reserved for one purpose alone, to serve as the outermost shape of the whole universe. So when Hephaestus forged twenty threelegged cauldrons to stand round the walls of his well built hall, he was really making the external form of the whole universe.

The external form or outermost shape of the universe was regarded by later thinkers as the region of the fixed stars. For this reason they said that the fifth regular solid, our ball, was adorned with the figures of living creatures. These figures were the constellations of the fixed stars, the groupings of the stars in the different regions of the heavens so as to form the shapes of earthly creatures, mythical beasts, and heroes and heroines. These were thought of as adornments to the heavens and they have their counterparts in the design of Hephaestus' cauldrons, the ornamental handles which he is about to affix to the cauldrons when Thetis arrives. He has made the handles and finished cutting the rivets with which to attach them to the cauldrons, an excellent moment to stop working and entertain his guest.

The fixed stars are fixed in relation to each other, unlike the planets which are always wandering about. But from the point of view of the earth the fixed stars are in constant motion and make one complete circuit of the heavens every day. That is why the cauldrons are described as having legs with wheels on them, so that they can skate around the outermost reaches of the cosmos. But it is notable that like the fixed stars they do not move independently of each other but all together in unison, and that they return to just where they started to the amazement of everyone who sees them. They run by themselves just as the later Greeks believed the heavenly bodies to be divine, self propelling intelligences whose superior reason is shown by the absolute predictability of their movements. And they run to a gathering of the gods. This is entirely appropriate for threelegged cauldrons since these tripods were among the most common votive offerings made to the gods. In this context, however, it may even be that the gods represent the planets, as they are certainly represented by planets in both the Egyptian and later Greek theology. If they do represent planets here, then the movement of the cauldrons among them may serve as an image of the ever changing relations between the planets and the constellations of the fixed stars. Understood in this way, these few lines about the tripods are a remarkably concise image of the heavenly bodies in motion.

The two accounts of Hephaestus at work which we have so far considered are alike in a number of ways. Both accounts are incidental to the main action of the story but both are significant of the god's real nature; the work which they describe is essential to his part as the divine craftsman of the universe. What Hephaestus made in Thetis' cave, and what he is making as Thetis arrives at his palace, appear at first to be the creations of a casual hour. When their real meaning is set against their casual appearance we are given an insight into the relations between the mortal and the immortal, the accidental and the essential. But there is one other feature which these two accounts of Hephaestus at work have in common and that is the apparent motive for the work. Why does he make what he makes in Thetis' cave? To pass the time and please his protectress. Why does he make the tripods? To amaze the gods by their running back and forth of themselves. On both accounts the motive for creation seems to be playfulness, a wish to delight. When we look at the fantastic display of the creatures under the sea or the intricate movements of the stars, we may feel that this is the truest and most profound reason that can be given for the creation of worlds. They are just wonderful toys and until we know it we have not begun to understand. This motive becomes even more impressive when we remember the enormous effort required for the forging of metals.

Let us return now to Hephaestus in his palace as he prepares himself to meet Thetis. He has put his tools away, sponged himself down and put on a tunic over his shaggy breast. Then he takes hold of his staff and begins to leave his workshop.

> Golden maidservants hastened to help their master. They looked like living girls and could not only speak and use their limbs but were endowed with intelligence in their minds and trained in handwork by the immortal gods. Supported by his toiling escort, Hephaestus made his clumsy approach to the place where Thetis was seated.

This passage raises more acutely a problem we passed over rather lightly in the case of the cauldrons. The cauldrons too moved of themselves, but apart from pointing out that in this they were like what they were taken to represent, the heavenly constellations, we left the question of their self propulsion unexamined. Now we have these maidservants, who are also inanimate and merely like living girls but who can move and speak and think and are skilled in the crafts. It is very striking that in this first account of the smith in the Greek tradition we should find him surrounded by contrivances sufficiently far advanced to stir the imagination of inventors today. Yet we must be careful not to assume that the cauldrons and the maidservants are machines, for all that they are made of metal. At the most there does seem to be a connection between their being capable of moving themselves and their being of gold. The maidservants are of gold as are the wheels of the tripods.

It is disconcerting that these maidservants should have all the faculties of human beings, including speech, intelligence and skill in the crafts, and yet be merely like living girls. The maidservants are just like us except that they are not alive, are made of gold and reside in Hephaestus' palace. These differences between ourselves and the maids compel us to ask the question 'If the maids are like us in every other respect, yet are not alive, what drives them?' Homer, I suspect, describes the maids as he does to force us to this question.

What makes the maids go? What makes them go has to be Hephaestus. How he does it is less clear but it is not by any mechanical agency. They are not empowered by engines but directly by the god. This is by no means an uncommon notion in Greek thought. God is frequently described as effortlessly mobilising the universe by the thought of his mind. Sometimes this power is attributed to the thunderbolt, the lightning brand, and this notion is still closer to the description of Hephaestus in the Iliad since Hephaestus works with fire and his name is sometimes used as a synonym of fire. Hephaestus is not merely the creator of worlds, he is also the power which animates them. He is the vital warmth, not merely physical but spiritual, which gives to creatures the powers of consciousness and of moving themselves. That is why his creations are always on the point of turning into living creatures. They are not only lifelike, they are actually alive as some of the ancient critics noted.

The cauldrons and the maids who are otherwise so human are animated by the fire of Hephaestus. But if this fire can make the maids go, may it not also be what makes us go? From a certain point of view we too may be considered dolls or puppets, empowered by a force which we cannot comprehend but which sustains us at every moment and in every faculty. Understood in this way, the maidservants are Homer's way of showing how we too are made and moved by Hephaestus, our most subtle faculties the work of his forge. Like the cauldrons the maidservants seem to move of themselves, but really they are activated and controlled from within. They only move intelligently in obedience to this and in its service. It is just in this appearance of free will that they most of all resemble us. Like them we could not move without a power put into us and continuously transmitted to us by an intelligent principle distinct from any and all of our moving parts. 'We who dwell on earth can do nothing of ourselves' wrote William Blake, and St Paul 'Not I but Christ in me'. This is the power in whom we live and move and have our being, and all the automata which we ourselves make imitate the creations of the mythical craftworker who made us. This view of human action is closely connected to the Greek view of inspiration. Even Homer's song is not his own, but is transmitted to him by the Muse.

Through a variety of symbols Homer shows how Hephaestus formed the whole of the universe, from the constellations to human beings and the forms of marine life. Furthermore, we have inferred that Hephaestus is the essential fire which animates all living things and gives them the powers of thought and movement. As with the other gods, so in the case of Hephaestus we have to do with a totality, a power or being which is truly universal, to be found everywhere and in everything. In this respect the study of the ancient Greek theology is overwhelming, since the understanding of each god or goddess seems to provide a complete explanation for everything. But though each divinity is a totality in this way, there are of course all the others as well. From one point of view each divinity is a unity which brings everything into a single scheme or system of understanding. But from another point of view each divinity is merely partial, one path of the understanding among many. The advantage of this plurality is that the Greek theology is able to distinguish very exactly between many different aspects of the divine. In the Christian theology, on the other hand, the creative power of God is not so easily distinguished from the eternity of God or from the divine justice and mercy.

But if Hephaestus is all these things, why then is he lame? To this question there have been a great many answers. Some say that he is lame because smiths work with their arms, not their legs, and it is true that Homer often describes Hephaestus as strong in the arm. Others say that he is lame because it was the practice to cripple slaves who knew how to work with metals to prevent their running away. They had to suffer this discrimination because they were more valuable than other slaves. Again, it is often said that Hephaestus is lame to demonstrate his inferiority to the other gods, because he is a worker among aristocrats and it would be improper for him to be on the same footing as his social betters. This is a very plausible answer because it is certainly true that the other gods find the sight of Hephaestus' hobbling about extremely amusing and they take no trouble to hide their amusement. He sometimes pours their wine and he built their palaces. But what is wrong with all these answers is that they do not recognise that Hephaestus really is a god. Instead they treat him as nothing more than a reflection of the human order.

In no passage of Homer is Hephaestus as lame as he is in this one. Here he is in his own palace after being seated at the forge, yet to go from one room to another he requires the assistance of a staff and toiling maidservants. Elsewhere in Homer's work he is described as circulating among the Olympians and pouring their wine in the palace of Zeus, and even in this passage he is described as nimble enough on his legs as he rises from the anvil. It is significant that his lameness should be so pronounced just when we are given the account of the maids. The maids represent human beings who are also automata governed by a divine intelligence. Likewise the human limbs are automata. Perhaps to emphasise this point, Homer pictures Hephaestus as making use of the automata he has made as a substitute for his own limbs. Hephaestus' dependence on the maids may convey the same message as the maids' likeness to humans. Both suggest that the living creature is an automaton manufactured and animated by Hephaestus in part and in whole. Once granted that the god can be described as having human form, there is no impiety in suggesting that that form is defective if the defect helps to reveal the true nature of the god.

At last Hephaestus makes his way to the room in which Thetis and Charis are sitting. He sits down close to Thetis, takes her hand and like Charis reproaches her for not coming to see them more often. He then begs Thetis to tell him how he can be of service to her. Thetis bursts into tears and tells him the sad story of her son Achilles, ending with the request that Hephaestus make Achilles new armour. Hephaestus immediately agrees, sorry only that he cannot do more for his guest. He stands up and immediately goes back to his workshop to begin making the new armour. He sets the bellows to the fire under the crucibles and tells them to start blowing. It is no surprise, after the wonders we have already seen, that these bellows should be fully automatic like the maidservants and the tripods. They respond directly to Hephaestus' commands, varying the force of their blast to the needs of each stage of the work. Thus our highly developed technology may be seen as the realisation of Homer's dream of Hephaestus.

Next Hephaestus throws bronze and some tin and gold and silver onto the fire and begins to make a shield, the most famous shield in history, five layers thick and covered with designs. First he represents

> earth, sky and sea, the tireless sun, the moon at the full and all the constellations with which the heavens are crowned, the Pleiades, the Hyades, the great Orion and the Bear...

After what we have seen, it is clear that the designs on the shield do not merely represent a universe already in existence. The making of the shield of Achilles gives Homer another chance to describe Hephaestus at the work of universal creation. The making of the shield is a metaphor for the making of the world, a metaphor no more and no less oblique than those which have preceded it. Under the one description of the shield Homer succeeds in combining his fullest account of the divine creation with a single incident in his story of the all too human Achilles.

After making the heavenly bodies, Hephaestus creates two cities. In the one marriages are being celebrated and a law suit is under way. The other is under siege by its enemies. The two cities represent the twin principles of love and hate, or peace and war, or civility and violence, the fundamental principles whose interactions govern the universe according to the later Greek philosopher Empedocles. In a battle scene next to the beleaguered city Hephaestus depicts Death, her cloak red with the blood of the fallen, and all in metal. He creates fields and streams, herds of cattle and flocks of sheep. In one place a bull is being attacked by two lions while the dogs of the cowherds bark but keep their distance. He depicts a vineyard, all of gold though the bunches of ripe grapes are black and the props that support the vines show up throughout in silver. All round the vineyard is a ditch made of blue enamel, and round that again is a fence of tin. We may guess that the blackness of the grapes is achieved by means of a sulphide of copper. He shows ploughmen ploughing a fallow field and receiving a drink to refresh them at the end of each furrow. The field they are ploughing is made of gold but it grows black behind them as a field does when it is being ploughed.

> Next the god depicted a dancing floor. Youths and marriageable maidens were dancing on it with their hands on one another's wrists, the girls in fine linen with lovely garlands on their heads, and the men in closely woven tunics showing the faint gleam of oil, and with daggers of gold hanging from their silver belts. Here they ran lightly round, circling as smoothly on their accomplished feet as the wheel of a potter when he sits and works it with his hands to see if it will spin; and there they ran in lines to meet each other. A large crowd stood round enjoying the delightful dance, with a minstrel among them singing divinely to the lyre, while a pair of acrobats, keeping time with his music, threw cartwheels in and out among the people.

Finally, right round the rim of the wonderful shield, he put the mighty stream of Ocean, the limit of the world.

It is a pleasure to think of Hephaestus as he bustles about his automatic forge, the bellows blowing directly in response to his spoken orders.

There he is, surrounded by his beautiful golden maidservants, and we must remember as we think of him that this description was composed some thirty centuries ago, at the very beginning of our era. The contemplation of such things restores to the metallurgy of every age something of that splendour and pride in work which invested all the ancient crafts and professions. For the Greeks and Romans Homer's account of Hephaestus ensured that metallurgy was exemplary among the crafts in this regard. Mining and metallurgy are prime sites for exploring that theory of work which is arguably the normal theory of work, the one found in the most enduring civilisations. According to this theory every kind of work is sacred as a manifestation of the divine nature. Every kind of work realises a different aspect of the divine.

PROMETHEUS AND ARACHNE

But if the imitation of the great Hephaestus conferred upon the human smith a certain glory, it brought with it some terrible responsibilities. The ancients revered the arts and crafts but they also feared them. This is the force of many of their stories, that the imitation of the divine creative powers was always extremely dangerous and to be hedged round by rituals of appeasement. This combined love and fear of technology has something in common with certain attitudes in our own time, especially in the case of mining. The ancient myth which most fully explored these twin emotions of love and fear was the story of Prometheus.

A very long time ago there was a great war between the gods, between Cronus who was king of the Gods and Zeus, his son, who wished to usurp his father's throne. The gods divided themselves between these two leaders. The older gods, or Titans, sided with Cronus while the younger gods supported the claims of Zeus. Eventually Zeus and his supporters were victorious and established themselves as the new rulers of Heaven. Cronus and his followers were banished and a new order began. At this time the human race was not as it is now, although it already existed. According to the poet Aeschylus in his famous play *Prometheus Bound,* at this time

> humans knew nothing of houses built of bricks to face the sun nor of woodwork, but lived underground like swarming ants in sunless caves.

Zeus, the new ruler in Heaven, decided to annihilate this race and make another. But Prometheus, a minor god and one of the few Titans to have fought on Zeus' side, took pity on the human race and wanted to help it. He secretly entered Hephaestus' workshop and stole some of the fire there. This he carefully placed in a hollow stalk of fennel, to hide it and protect it. In this manner he brought fire down to the earth and gave it to the human race.

Not only did he give fire to humans, he taught them letters, astronomy, mathematics, the domestication of animals, medicine and augury. In short, every art and science possessed by humans came from Prometheus. In Aeschylus' play Prometheus finishes the long catalogue of his gifts to humans:

> and then there were the benefits to humans hidden beneath the earth. Bronze, iron, silver, gold, who would claim to have discovered them before me?

Prometheus' theft and his saving of the human race from destruction by giving it divine powers infuriated Zeus who visited the most terrible punishment on him. Aeschylus' play is set in the mountains of the Caucasus. The play opens when Hephaestus and two lesser gods, Strength and Violence, lead Prometheus on to the stage and proceed to fasten him with bonds of bronze to the side of a rock. There he is to be left at Zeus' pleasure, to suffer the extremes of cold and heat by night and day. Hephaestus, as usual, is sympathetically portrayed. He is at odds with the cruelty of his assistants. He deeply regrets having to do what he is doing to Prometheus, and declares that he hates the craft which qualifies him to do it. Throughout this opening scene Prometheus is silent, while Strength urges Hephaestus to tighten and add to his bonds and Hephaestus miserably complies. The climax comes when Strength insists that Hephaestus drive a wedge straight through the immortal Prometheus' chest. Groaning in sympathy Hephaestus does this too.

Hephaestus' hatred of the craft which Zeus has compelled him to use against Prometheus is very different from the attitude to his work which Homer describes. But this hatred is typical of a play and a myth in which the value of the crafts is radically at issue. Prometheus is a philanthropist who saves the human race and gives them almost everything that makes human life worthwhile. But this is not the only sense we have of him in the play. Perhaps because almost all the characters are gods or goddesses, we are frequently made to feel that Prometheus' generosity to mortals has overstepped the mark, especially taken in conjunction with his lack of care for himself. Prometheus obtained the fire which he gave to mortals by theft and all his gifts were given us contrary to the will of Zeus.

The story of Prometheus suggests that the arts and sciences which he stole from the gods have placed humankind in a condition of disequilibrium. The story captures a feeling about human life which is common to many cultures, that we are an uneasy mixture of god and beast, a mixture of elements which should never have been combined. The description of Hephaestus at work in the *Iliad* casts a special lustre over the work of human smiths and it is meant to do so. In traditional societies all activities of this kind are sanctioned by divine patronage. But this relationship with the divine is as much a matter of dread as of pride. The story of Prometheus and Aeschylus' version of it make us feel this dread very acutely. There is no suggestion that these powers of the gods are too great for humans to handle, that our technology may run away with us beyond our control. The suggestion is simply that these arts and sciences of the gods are worth much more than we are, and that we had better be conscious of it. To forget this is to be guilty of the tragic insolence which brings catastrophe upon Prometheus.

The same point is made by the story of the weaver Arachne. She was brilliant at her loom and very quickly earned a reputation as the finest weaver of all time, famous as much for the grace of her movements as for her finished work. Now of course this skill must in some sense have been taught to her by Athene, the goddess of weaving and many other crafts. But Arachne refused to concede this. So Athene decided to visit her, disguised as an old women, and pleaded with her to acknowledge Athene's superiority in weaving. Athene tells her that she is free to claim to be the best of mortal weavers but must not usurp the prerogatives due to divine beings. Arachne rudely dismisses the old woman, who immediately transforms herself back to Athene and challenges Arachne to a contest. Both competitors produce marvellous tapestries, Arachne weaving the figures of all those mortals unjustly treated by the Gods. When both have finished, Athene can find no flaw in Arachne's work and goes wild with fury. She tears Arachne's tapestry to pieces, then hits Arachne on the head with her shuttle. Arachne refuses to put up with this and hangs herself. At last Athene relents. Arachne, she says, may go on living but must remain suspended. Then she sprinkles poison on Arachne which turns her into a spider. Her arms are joined to her sides and those nimble fingers turn into spider's legs as the rest of her body shrinks.

The primary aim of this story is to explain the genesis of the spider. Like many such myths, especially among the Aboriginal people of Australia, it shows how the other creatures in nature were originally human and are therefore still our kindred. But as is also very common, the story turns on the ethical doctrine that the arts and crafts come to us from the gods and goddesses and must be recognised to do so. The point is not that Athene cheated in refusing to acknowledge the perfection of Arachne's tapestry. It may even be that Arachne was capable of such perfection just because Athene, who is the source of the weaver's inspiration, was working right beside her. The point is that Arachne's skill is wholly the result of Athene's teaching. Just as the golden maidservants of Hephaestus were made and empowered by him, automata wholly dependent on the divine craftsman, so Arachne's skill is nothing other than Athene's working through her. The traditional mind cannot make much sense of the proposition that Arachne might have managed what she achieved without Athene, since without Athene there would have been no one there to do it. Arachne was suffering from a delusion. But it is a delusion to which clever people are particularly susceptible and which brings disastrous consequences upon them. An identical story is told of the satyr Marsyas who challenged Apollo, the god of music, to a music contest. When he lost, Apollo flayed him alive.

The underlying assumption of these three stories is the same. When Prometheus stole the fire from Hephaestus' workshop and gave it to the human race, we acquired something to which we have no right. Though these skills and sciences are the basis of our civilisation, they are not ours by right but by the grace of the gods. That the gods should condescend to us in this way imposes upon us the duty to acknowledge their preemininence in all these activities, and to recognise at all times their graciousness to us. Such a response has two results. Firstly it makes it very difficult for people to take these arts for granted. The duty to acknowledge the gods' part in them enforces and reinforces the sense that they are somehow miraculous. And secondly the belief of the craftworker that it is not the human self but a divine power which does the work, helps to bring about that denial and destruction of the egoic selfhood which is the goal of all traditions and traditional forms of work.

This is as true of mining and metallurgy as it is of the other crafts and professions. Over half the mines of which we know the names in ancient Greece were named after gods or religious personages. It is probable therefore that the ancient Greek miners and metal workers were as devout in their way as many of their Christian counterparts, and in Christian times these professions have often been noted for their piety. The piety of these people did not stem merely from fear of their dangerous occupation but from the recognition that what they won from the earth and worked, was, first and last, the gift of Heaven.

PLATO'S THEORY OF CRAFT

The Greeks respected the arts and crafts to an exceptional degree, but this respect was not unqualified. Homer, for example, explicitly subordinated the creative powers of the divine to other powers. Hephaestus the smith is an Olympian but he was not king on Olympus, nor does he take much part in the councils of the gods. Zeus is the king of the gods and Zeus is no craftsman. In this way Homer made it clear how the different powers or aspects of the divine were to be ranked. This subordination of the creative powers is much more obvious in the Greek theology than it is in the Jewish or Christian. The Old Testament begins with an account of the divine creation, and this aspect of the divine nature has been much more central to the western understanding of God since the beginning of the Christian era.

In the Greek theology the emphasis was different. The Greek theologians certainly believed that the physical universe depended for its existence on the divine creative powers, represented by Hephaestus. But this mattered much less to them than the related principle that the divine nature in itself is utterly self sufficient, depending on nothing whatsoever. It is what it is by virtue of its own nature, absolute, unchanging and eternal. In this way the divine is quite different from what is created. According to these theologians, Zeus and other Olympians represent these metaphysical aspects of the divine, and to these Olympians Hephaestus is inferior because of his involvement with what is less than absolutely self sufficient. This may be another reason why Hephaestus is presented as lame, to distinguish him from those higher aspects of the divine nature which are signified by the able bodied gods and goddesses. In traditional societies, for the same reason, the purely contemplative life is valued above the productive and creative lives. And this was true even of the Christian world, despite its greater emphasis on the creative powers of God.

In these ways, then, the divine creative powers are less than supreme. They are less than supreme because they produce or act upon material things which are not self sufficient but depend for their existence upon what is self sufficient. The pure contemplation of what is self sufficient, to no other end than for the sake of that contemplation, and in which the one who contemplates is absorbed into the object of contemplation, this is the highest and best of all lives. The crafts and professions are not purely contemplative because they act upon or produce material things. But this is not to say that the crafts and professions are not contemplative at all. On the contrary, all work proceeds from contemplation to action and is made up of these two theoretically distinct parts. Of these two parts the contemplative is traditionally regarded as the better, while the active which proceeds from it and is directed by it, is the worse. Work is sacred because it is contemplative.

The belief that work requires contemplation is common, I dare say, to all traditional societies, including those such as the Australian Aboriginal in which there is no simple division of labour. The belief is particularly strong in the civilisations of India and China. In medieval Europe, a time and place where the arts and crafts flourished, this doctrine of contemplation was summed up in the Latin 'Laborare est orare', to work is to pray. This prayer which is work is not added to the work, not an incidental activity carried on at the same time as the work, but the essence of the work itself. In this way all work leads to the development or unfolding of the spirit in the traditional worker. The most complete exponent of this doctrine in the range of the western traditions is the Greek philosopher Plato, from whom this and many other insights were adopted and adapted by the Christians.

Plato argued that just as a painter looks at the object which he is painting in order to capture its appearance in his work, so the craftsman constantly refers to the idea of the thing which he is making in order to make it. The carpenter must have in his mind's eye the idea of the bed or table which he is working on. Likewise the smith contemplates as he works the idea of the axe or cauldron which is being made. Of course he might just copy some other smith's axe or cauldron but in that case he is not fully a smith. It is precisely this capacity to entertain the idea of the thing to be made which confers upon him his expertise. This idea is not, according to Plato, different for each thing which the smith makes. All axes are copies of the one idea of the axe, and all cauldrons of the one idea of the cauldron. The one idea of each artifact inspires many different versions of itself just as the one landscape can inspire many different pictures.

For Plato the idea of the axe is not something we learn from the axes we come across. It is not a concept derived from our physical experience. On the contrary, it is by looking at the invisible idea of the axe that smiths bring the axes we know into existence. In this way the idea of the axe comes first and the axes in the world come from it. Where then does the idea of the axe come from? According to Plato the idea of the axe is not in this world at all. It belongs to another order of reality, a world of invisible ideas which do not change like the things in this world but which are the originals of all the things we know here. This world of ideas includes not only the originals of all the artifacts in this world but the originals of all the natural kinds and species as well.

Imperceptible except to the eye of the mind, these ideas of natural things and artifacts do not exist in the world of time and space. But they are, according to Plato, the real causes of everything in time and space. He compares them to objects carried between a fire and the back wall of a cave; the shadows which these objects cast upon the wall are images of the objects just as the things in this world are images of the ideas. Like the objects which cast the shadows, the ideas are really much more substantial than things in the world, though to the untrained and underdeveloped mind they appear somewhat vague at first and difficult to grasp. In the same way what the visionary artist conceives in his imagination is much stronger than any version he can make of it. Often he must create many such versions in order to capture even a part of its magnificence.

This concentration upon the invisible idea is the contemplative and superior part of any work, while the embodying or realising of the idea in the physical world is the active and inferior. It follows from this that the proper beginning to any work is contemplation of the idea which the work is to realise, a contemplation which should then be carried into the actual labour. In many cultures this preliminary contemplation is ritualised, as in the fasting of the icon painter or the meditation upon 'the dreamtime' of the Aboriginal bark painter. The making of something, the bringing of something into the world, is the bringing of what is invisible into the realm of the visible. It is a bridging of the two worlds, a uniting of the ideal and the temporal, in which the maker acts as a channel between the world of the spirit and this world. According to the Greek theologians, the twenty threelegged cauldrons of Hephaestus were the outermost shape of the cosmos. Hephaestus modelled the cosmos after one of the five regular solids. This is the mathematical idea which he realised in making this world.

In traditional societies these ideas of the artifacts and the ideas of nature are regarded as belonging to exactly the same order of reality. The idea of the axe is as much part of the 'natural' order as the idea of the tree. In these societies there is no sense that the world of human artifacts stands outside or threatens the natural world. Human weaving, as we have seen, is the origin of the spider's web. Similarly those arts and crafts which foster or develop natural resources are regarded as contemplative in exactly the same way as the crafts which produce distinctively human artifacts. Just as the smith must fix his mind's eye on the idea of the axe or cauldron, so the shepherd or sheep breeder must contemplate the idea of the sheep if he is to preserve and improve the creatures in his care. The miner must know the idea of the metal he seeks if he is to find and free it from the earth, no less than the metalworker must know the idea of whatever it is he will make from that metal.

This contemplation of the idea is called in some cultures the free act of contemplation. It is contrasted with what is called the servile act of manufacture. The mere physical labour of doing what has to be done to realise the idea in the world stands to the contemplation of the idea as the slave to the free. It is, comparatively, a constraint upon the spirit of the worker to have to use hands, while in contemplation the spirit is free of the limitations of time and space as it is absorbed into what is beyond them. It is also for this reason, perhaps, that Hephaestus is both divine and lame, powerful in the spirit but at the same time limited in body. Another way of making the same point is to say that the art is far stronger in the artist than in his works, since the idea which the artist contemplates in the spirit is far greater than any embodiment of it. In this way the practice of traditional work was regarded as a means of building up the spiritual powers of the worker by demanding an ever greater capacity for the contemplation of the invisible. The traditional crafts were at least as much concerned with the development of those who practised them as with the material products of their work. The first reason Plato gives for the division of labour in his Republic is that it develops the natural propensities of his citizens, each of whom has a talent for a particular kind of work. The second reason he gives is that it is a more efficient way of producing better goods if labour is divided.

If Homer is the poet of the crafts, Plato is their philosopher. But we do not find in the works of Plato any extended account of a single craft to compare with Homer's account of the smith. Plato refers to individual crafts as examples or metaphors for the purpose of developing large scale theories. There can be no doubt, however, that he too was susceptible to the special beauty of metals as the following description of the afterlife makes clear. In this passage Plato is describing how the world looks to someone who has managed to escape from the limiting conditions of a mortal life. He has just suggested that the earth is really much bigger than anyone imagines and that we who think we live on its surface in fact inhabit deep hollows or pits into which impurities have gathered. The difference between the earth we know and the real earth is quite as great as that between the world beneath the sea and the world above it. If one could escape from the hollows and rise above the mist which is our normal atmosphere, a world would appear as startlingly different and pure as the one a fish first sees which pokes its head above the surface of the sea.

> The real earth, viewed from above, is supposed to look like one of those balls made of twelve pieces of skin, variegated and marked out in different colours, of which the colours which we know are only limited samples, like the paints which artists use; but there the whole earth is made up of such colours, and others far brighter and purer still. One section is a marvellously beautiful purple, and another is golden; all that is white of it is whiter than chalk or snow; and the rest is similarly made up of the other colours, still more and lovelier than those which we have seen. Even these very hollows in the earth, full of water and air, assume a kind of colour as they gleam amid the different hues around them, so that there appears to be one continuous surface of varied colours. The trees and flowers and fruits which grow upon this earth are proportionately beautiful. The mountains too and the stones have a proportionate smoothness and transparency, and their colours are lovelier. The pebbles which are so highly prized in our world, the jaspers and rubies and emeralds and the rest, are fragments of these stones, but there everything is as beautiful as they are, or better still. This is because the stones there are in their natural state, not damaged by decay and corroded by salt water as ours are by the sediment which has collected here, and which causes disfigure-

ment and disease to stones and earth, and animals and plants as well. The earth itself is adorned not only with all these stones but also with gold and silver and the other metals, for many rich veins of them occur in plain view in all parts of the earth, so that to see them is a sight for the eyes of the blessed.

This passage works at several different levels of meaning simultaneously. It purports to be a description of this earth of ours as it really is, but we cannot, I think, be expected to take it seriously at this level. Are we really to suppose that the surface of our earth is so utterly unlike what we know of it? It is more reasonable to assume that Plato has chosen to frame his description in this way as an explanatory device rather than as serious geography. When we read at the beginning of the passage that the earth is like one of those balls made of twelve pieces of skin, we are of course reminded of the way in which the ancient commentators explained the twenty threelegged cauldrons which Hephaestus is making when Thetis visits his palace in Homer's Iliad. The twenty threelegged figures were taken to be the twenty points or solid angles, at each of which three edges join, on the surface of just such a ball as Plato is describing. According to the commentators, Homer intended this figure to represent the outermost shape of the cosmos. Elsewhere Plato himself uses this twelve sided ball as the figure of the cosmos as a whole, and this suggests that the real earth which he is describing here is the whole physical universe and not just the terrestrial globe. On this interpretation it makes very good sense to say that by far the greater part of this earth projects above the atmosphere which we inhabit. Furthermore, to describe this larger world as the place of the afterlife is comparable to our own expression 'the astral plane' to denote the same thing.

Beyond even this level of interpretation is another, according to which this real earth which Plato describes is not a material world at all. On this view the glorious world which Plato depicts is a metaphor for the world of the ideas, the home of the pure intelligence beyond time and space. Here are the ideas not only of the artifacts and of the natural species, but the mathematical and moral ideas as well. To people who are not accustomed to the self conscious contemplation of ideas it makes little sense to describe them in their own terms. Instead Plato develops an elaborate metaphor to express something of this other world by the use of physical analogies. The things which he chooses to stand for the ideas in these analogies are the precious stones and the metals. By their purity, durability and brilliance they stand to the rest of the physical creation as ideas stand to all physical things. The precious stones and the metals are the proper symbols of the invisible ideas in the material realm. As we shall see, this is one of their most constant functions in the history of western thought.

Chapter Three

THE BIBLICAL TRADITION

Then the Lord spoke to Moses, saying "You shall also make a lampstand of pure gold; the lampstand shall be of hammered work. Its shaft, its branches, its bowls, its ornamental knobs, and flowers shall be of one piece. And six branches shall come out of its sides; three branches of the lampstand out of one side, and three branches of the lampstand out of the other side. Three bowls shall be made like almond blossoms on one branch, with an ornamental knob and a flower, and three bowls made like almond blossoms on the other branch, with an ornamental knob and a flower... Their knobs and their branches shall be of one piece; all of it shall be one hammered piece of pure gold. You shall make seven lamps for it, and shall arrange its lamps so that they give light in front of it. And its wicktrimmers and their trays shall be of pure gold. It shall be made of a talent of pure gold, with all these utensils. And see to it that you make them according to the pattern which was shown you on the mountain."

Exodus

GOD WAS A SMITH for the Jews as well as for the Greeks. It is well known that when Moses went to the top of Mount Sinai he received the ten commandments, written on stone tablets by Jehovah. It is less well known that on this same occasion Moses received instructions on how to make the ark of the covenant and the tabernacle in which to house the commandments. These instructions were given by Jehovah in the fullest detail and they show that the God of the Jews and Christians was no less a designer in metals than the Greek Hephaestus. Metallurgy is the exemplary craft of the Jewish tradition as it is of the Greek. Similarly it is well known that the promised land to which Moses led the children of Israel was a land flowing with milk and honey. But Moses also described the promised land as

...a land where the stones are iron, where the hills may be quarried for copper.

In turning from the Greeks to the Jews, we must not confuse the two traditions with each other. This is the easier to do when we consider first one and then the other as we are doing here. In both traditions there is a patriarch who establishes for his people the theological, ritual and moral principles which all later generations venerated and embellished. For the Greeks this patriarch was Homer, for the Jews Moses. The principles of these two patriarchs have certain elements in common, and throughout Western history there has been a strong tendency to conflate them so as to create a single relogous order. The task has been to turn Homer into Moses or Moses into Homer. On the whole the inclination has been to turn Homer into Moses since in this way the Christian view is given precedence over the pagan. The early Church fathers sometimes argued that Homer had learnt all he knew from Moses, who was the fountain of all Western thought, the first wise man from whom all the others had learnt. Modern criticism has served both Homer and Moses in exactly the same way. By arguing that the works ascribed to them are composed of passages from many different historical periods, modern critics have cast doubt on the existence of any single author responsible either for the Iliad and the Odyssey or for the first five books of the Bible.

These critics have suggested that there are passages in the Mosaic books which were composed later than anything in the Iliad and the Odyssey. One of these passages occurs, indeed, in the instructions for the tabernacle. This means that the Jews may have been influenced by the Greeks of Homer's time and later in the formation of their scriptures. As far as the dating of their scriptures goes, this is no less possible than that the Greeks were influenced by the Jews. We must also remember that both the Greeks and the Jews were traders and that the Greeks were accomplished sailors from early times. It is by no means unlikely that such similarities as there are between the two religions are the result of a direct transmission, one way or the other or both, between the two nations.

But there are other possible explanations for these similarities. One is that religious traditions always have much in common just because they are religious traditions. It has been said that there is no people among whom the Lord has not left a witness to himself, a statement which supposes that the religious point of view is common to all human societies, even those which are quite isolated from others and where there can be no question of transmission. Even in societies such as these, it has been said, the religious sensibility is fully developed, and recognisable forms of worship are in use. If this is so, we do not need to explain the similarities between the Greek and Jewish traditions by supposing a direct transmission between them, or by the inheritance of common doctrines from a single source. We may suppose instead that the doctrines arose spontaneously in both places, and were developed quite independently.

GREEKS, JEWS AND EGYPTIANS

But the fact remains that the smith god occupies a more prominent position in the Greek and Jewish traditions than he does in others. In the eastern Mediterranean region there was one other early civilisation in which the smith god played an important part, and that was Egypt. Here his name was Ptah. Egypt exercised a very great influence over both the Greek and the Jewish cultures and it is quite possible that they both inherited the doctrine of the smith god from this common source. Herodotus, the first Greek historian whose work has survived, wrote that the Greeks acquired all their gods and goddesses from the Egyptians except Poseidon, Castor and Pollux. He also wrote that Homer was the religious teacher of the Greeks. Putting these two statements together, we infer that Homer taught the theology of Egypt. As for Moses, he was brought up as an Egyptian noble and was educated in the sciences of Egypt. Given all this and Ptah's intrinsic relevance to our topic, we must look at him more closely.

The god Ptah was one of the first and greatest of the Egyptian pantheon. He was worshipped from the second dynasty of the Old Kingdom down to Ptolemaic and Roman times, for well over three millennia. He was called 'the very great god who came into being in the earliest time', 'father of beginnings and creator of the eggs of sun and moon', 'the god who created his own image, who fashioned his own body'. One of the texts in the Theban recension of the Egyptian *Book of the Dead* concerns a man called Nebseni who was a scribe in the temple of Ptah. Ptah's greatest temple was at Memphis and this city was itself called the House of the Double of Ptah. In Thebes, the city of Amen Ra who was king of the gods, papyri have been found in which Ptah is made to possess all the attributes of all the great gods of Egypt. Not only was Ptah worshipped at Memphis and Thebes, he was worshipped also at Heliopolis, and this may be relevant to the question of how Judaism and the Egyptian religion were connected since Moses is sometimes supposed to have been educated by the priests of Heliopolis.

There are several different ways of representing Ptah in Egyptian art. In the vignettes of the *Book of the Dead* he is shown standing in a close fitting garment from which his two arms project. He wears a beard, is holding a sceptre and sometimes a flail, and he is standing under the canopy of a shrine. Elsewhere he is shown as seated, making the egg of the world on a potter's wheel which he works with his foot. He is almost naked but wears a close fitting cap and a uraeus. The meaning proposed for his name by modern scholars is 'sculptor' or 'engraver' and it is clear that he was the chief god of all the handcrafts and of all workers in metal, stone and clay. Hephaestus too is sometimes a potter, as when he makes the beautiful Pandora, the woman who brought all troubles to the human race.

From the *Book of the Dead* we learn that Ptah was the great artificer in metals, at once smelter, caster and sculptor, as well as the master architect and designer of everything which exists in the world. According to Wallace Budge, the Greeks and Romans rightly identified one form of him with Hephaestus and Vulcan. Elsewhere, however, he is described as carrying out the commands of Thoth in the making of the world, so that Thoth is sometimes thought to be a name for the intelligence of Ptah. Thoth had Maat as his female counterpart, who represents justice and sometimes the planet Venus. Insofar as Ptah assimilates Thoth to himself, he also acquires Maat and this may serve as a parallel to the marriage of Hephaestus and Aphrodite in the *Odyssey.*

Ptah created the world in accordance with the commands of Thoth, and he was particularly concerned with the construction of the heavens and the earth. We have already heard him described as creator of the eggs of sun and moon and pictured as a potter making the egg of the world. Ptah beat out the large rectangular slab of iron which formed the floor of heaven and the roof of the sky, and he and his assistants made the props which held it in place. In the *Book of the Dead* he is said to have covered his sky with crystal. In these descriptions he is concerned with the creation of the outermost limits of the universe and in some of them he is represented as a smith. This is comparable to Homer's account of Hephaestus in the *Iliad* as the maker of the twelve sided ball which is the figure of the universe as a whole. In Ptah's crystalline sky we have a parallel to the stars on the shield of Achilles and perhaps to the river Ocean which was the outermost circumference of the shield. In later Greek tradition the universe was often compared to an egg, of which the membrane next to the shell represented the crystalline sphere.

Ptah was closely connected with several other gods which he seems more or less to have assimilated as he assimilated Thoth. In association with Seker Ptah played a part in the ritual of the Hennu boat, the boat in which the sun sails over the sky in the second half of his daily journey. Once again we find Ptah in the context of the heavenly movements, in this case with the sun during the hotter part of the day. Ptah was associated with Khnemu in the making of the world. Khnemu made the humans and the animals while Ptah made heaven and earth. From a time much closer to Moses and Homer comes a prayer addressed to Ptah Tenen. This is the prayer found in the temple of Amen Ra at Thebes which ascribes to Ptah all the powers of the greatest Egyptian gods. Here the sun and moon are called the eyes of Ptah Tenen who knitted together the earth, who stretched out the heavens, who moulded gods and men, earth and sea. Finally mention must be made of his principal female counterpart, the goddess Sekhet who like the Hennu boat is usually identified with the west. Sekhet is the goddess of blazing fire and she is often just called 'Flame'. She is an appropriate consort for the god of the kiln and crucible.

Memphis, the ancient northern capital and major centre for the worship of Ptah, probably housed the most important goldsmiths' quarter in ancient Egypt. There is no evidence for the existence of a goldsmiths' guild but it is probable that the craft remained in the hands of a limited number of families working in one or two places, and was handed on from father to son for generations. It is certain that both the designers and executors of metal work were highly regarded in ancient Egypt and were often influential enough to build themselves large tombs. One goldsmith was appointed cup bearer to the king himself. It is likely that accomplished metal workers from elsewhere were attracted to Memphis to work in shops under royal or temple patronage, and that this concentration of talent persisted for a very long time. Memphis was certainly an important centre for metal workers in medieval times, and the tradition seems to have survived until quite recently in the metal workers of the Cairo bazaars.

GENESIS AND EXODUS

The cult of Ptah and his worship at Memphis account for only one of four distinct theories of the creation to be found in the Egyptian theology. Similarly the stories of Hephaestus in the *Iliad* and the *Odyssey* do not by any means exhaust the speculations of the early Greeks concerning the origin of the universe. The first book of the Bible also offers a number of different accounts of how the world came to be. In the first chapter of Genesis the creation of the world is the work of a divine craftsman. At the very beginning of the creative process there is a formless void and darkness, and then the spirit of God moves on the face of the waters. These phases correspond to the prayer and contemplation outlined at the end of the last chapter, the contemplation which must precede action in the making of anything. Then God said 'Let there be light' and by the mere act of his will there was light.

In these terms, according to the first chapter of Genesis, the whole world was created in six days. There is no indication that the various stages of the creation required anything more of God than that he affirmed them to be. It seems that the declaration was enough in itself to bring what was declared into existence. This way of putting the matter does not merely give precedence to contemplation over action, it seems to do away with the need for action almost entirely. In its sublimity it far outstrips the more material accounts of the creation by Hephaestus and Ptah, who have to work with their hands in order to create.

But at the beginning of the second chapter of Genesis, in a passage which seems to summarise the preceding chapter in its first words, we are told that God formed man of the dust of the ground and breathed into his nostrils the breath of life. This sounds more like the Greek and Egyptian stories, even though this manufacture is here restricted to the making of man and the animals. It does not apply to the creation of the earth nor of the heavens. God set this first man in a garden which he had prepared for him. In this garden the first man cooperated with the divine plan by tending and keeping the plants and trees. In this garden sprang a river which divided into four other rivers. The first of these four rivers surrounded the land of Havilah where there was gold and beryl. This is the first reference to metal in the Bible. The gold of Havilah, we are told, was good gold. Its connection with the garden of Eden suggests that for the Jews as for many other traditional peoples the paradisal state of our first parents was a golden age.

There is another quite different account of the creation in Genesis some chapters further on, and that is the story of Noah. If the creation of the universe in the first two chapters of Genesis was in part a matter of artifice, the work of a divine craftsman, the story of Noah presents us with an organic account. The story goes that within a few generations of Adam and Eve the human race had proliferated and become corrupt. So evil were its ways that God determined to destroy it, and he did so by covering the earth with a flood. He chose, however, to save Noah who had remained faithful to him by means of an ark or boat which he instructed Noah to build. Onto this boat Noah was told to take males and females of every species. The flood comes; all the other creatures on earth perish; the ark finally makes landfall; and Noah's family and the animals with them repopulate the world.

This story has close parallels with the Egyptian creation myths and with the story of Deucalion in Greek and Roman literature. The ark and the place where it lands are, as it were, the centre of the universe, the first point from which the universe expands outwards in all directions after the pattern of a seed or embryo. Of course, all the creatures in the ark existed before the flood, but the waters of the flood also symbolise the primordial waters out of which the first creation arose. In the midst of this nonentity the first point appeared, symbolised by the mountain top and the ark. Through the expansion of this point in all six directions the universe as a whole and space itself were brought into being. We need think only of the visual effect of the waters receding from Mount Ararat, or from Mount Parnassus in the Greek story, to gain a clear impression of the organic creation. The stories of Noah and Deucalion show us how the divine craftsman was not the only creative principle. Mythological accounts of the world as artifact coexisted with quite different ways of explaining how it came to be. Nor is it easy always to keep these different accounts of the creation distinct from each other. For example, from another point of view the garden of Eden may itself be regarded as an organic centre like Noah's ark, from which the rest of the creation was sprung.

Between these two accounts of the creation in Genesis appears the first reference to the working of metals in the Bible. At the end of the fourth chapter we are told that Lamech, the son of Methusael and the great great great grandson of Cain, had two wives and three sons. By his first wife, Adah, he fathered Jabal and Jubal; by the second, Zillah, he fathered Tubalcain:

> an instructor of every craftsman in bronze and iron.

In the Jewish tradition the name Tubalcain is closely connected with the name of Cain. The prefix is related in meaning to the word for spices, which are used to refine and improve the taste of food. Hence the name Tubalcain is taken to mean a person who refines or improves on the work of Cain. Cain was the first son of Adam and Eve and a skilful cultivator who tilled the ground. The Bible says that he built the first city. He also murdered his brother Abel. His descendant Tubalcain, the first metal worker, may therefore be considered to have improved on the work of his ancestor either by developing the instruments of agriculture and architecture, or by providing weapons for murderers.

Jewish thinkers have taken it that Tubalcain refined the art of murder. Probably they have done so because he appeared just before the flood, at a time when God was appalled at the wickedness of mankind. In accordance with this tradition the Renaissance philosopher Giordano Bruno included Tubalcain in his catalogue of inventors as contributing to the science of the battlefield. Against this must be put the tradition that Tubalcain's half brothers, Jubal and Jabal, did not invent destructive arts but benign ones. Jubal was traditionally the inventor of the harp and organ, and Jabal was the inventor of tents and herding. The simultaneous appearance of metal working and certain kinds of music inaugurated a connection between these arts which has lasted a very long time: foundry bands and the harmonious blacksmith. When we add to these arts those of the nomadic tent dwellers we have the components of the itinerant tinkers and the gypsies.

In the English versions of the Old Testament there are some verses concerning Lamech, Tubalcain's father, immediately after this brief account of Tubalcain. These verses read:

> Then Lamech said to his wives: "Adah and Zillah, hear my voice; O wives of Lamech, listen to my speech! For I have killed a man for wounding me, even a young man for hurting me.

> If Cain shall be avenged sevenfold, then Lamech seventy sevenfold."

After this the narrative resumes with Adam and the birth of Seth. As it stands, therefore, there is very little indication in the text to show what Lamech meant by this speech to his two wives. The fact that these lines immediately follow the account of Tubalcain and his inventions may have suggested that Lamech's killings had something to do with those inventions. From this too may have arisen the tradition that Tubalcain refined the arts of murder.

The Rabbinical tradition takes the speech of Lamech quite differently from the English translations. We must understand that Lamech was blind and his son Tubalcain used to act as his guide. They spent time together guarding their herds and flocks, Lamech with his great bow and Tubalcain as watchman. Their forefather Cain, the accursed, is still alive at this time. These patriarchs before the time of Noah were very long lived, in some cases for nearly a thousand years. Cain comes to Lamech's wives and asks where Lamech is. Lamech's wives tell Cain where he can find Lamech and Cain goes where they tell him. Tubalcain sees somebody or some animal approaching and warns his father who shoots an arrow. In this way Cain is killed. Cain was protected by God's vow that anyone who killed him would suffer sevenfold. It is because he has killed Cain that Lamech concludes his speech to his wives by referring to him.

When Lamech and Tubalcain discover that Lamech's arrow has killed Cain, Lamech is distraught. In his despair he strikes his hands together and, unable to see what he is doing, strikes Tubalcain's head between them. In this way he kills Tubalcain immediately after killing Cain, thus fulfilling part of God's vow that anyone who killed Cain would suffer sevenfold. Lamech's wives now leave him, and he tries to make peace with them by saying 'Hear my voice'. The Rabbis state that this phrase means 'Obey me and return to me'. According to the Rabbis, Lamech continues 'For the man I slew, was he slain by my wounding? And the young man I slew, was he slain by my blow?' Lamech's defence is that he did not intend to kill either Cain or Tubalcain, and therefore the wound and the blow by which he killed them were not properly his. He concludes with a threat. Just as anyone who killed Cain was to suffer a sevenfold vengeance, so anyone who injures Lamech will suffer seventy sevenfold, according to Lamech. The Talmud tells us how Lamech's wives responded to this threat. On Adam's advice they returned to him but could bear him no more children.

This story brings Cain and his namesake Tubalcain into much closer proximity but it does not demonstrate a connection between Tubalcain's inventions and the instruments of murder. Nonetheless the connection with Cain is perhaps enough to explain the negative Rabbinical tradition concerning Tubalcain. Cain murdered Abel in a fit of jealousy because God had accepted Abel's sacrifice of sheep and had rejected his own vegetable offering. One reason often given for this rejection was that God wanted his people to be nomads, but Cain had produced his offering by tilling the ground. On this view the nomadic peoples are much closer to the divine than are the settled because they are less attached to the things of the world. Agriculture requires settlement and we note that Cain built the first city. But Tubalcain went further. In developing the arts of metal working he was dealing with the most fixed and static of all the elements in creation. Of course, nomadic peoples in many parts of the world have developed metal working techniques, and we have already considered the itinerant tinkers. But the fact remains that these techniques are usually more highly developed among settled peoples.

Tubalcain and Prometheus are the archetypal teachers of metal working at the human level in the Jewish and Greek traditions. It is very noticeable that both should have led unfortunate lives. The two are unfortunate in different ways and for different reasons but they are both tragic figures. In both traditions the teachers of metal working come from a period of human history very different from our own, when the divine and mortal were much more closely related, when people lived for a millennium or learnt directly from the gods. That metal working should be singled out for special mention establishes it at once as central to the religious traditions of both these peoples, and the traditions of these two peoples are the foundations of our own religious understanding. Some would even see a connection between the names of Tubalcain and Vulcan, the Roman Hephaestus.

According to tradition the author of Genesis was both the author and the hero of Exodus. The story of how Moses brought the Israelites out of Egypt and their forty years in the wilderness is unsurpassed as an account of leadership. Out of that confused and turbulent mass of fugitives from slavery, Moses created a single people which has maintained his teachings as a living religion for thirty centuries. The account which Exodus gives of the first struggles of this people remains the founding scripture of a great nation. The commandments, rituals and ritual vessels which were then established for the first time are still the sacred core of Judaism today. In that desert and with these people Moses received directly from God the inspiration for the metallurgical wonders which are the focus of Jewish worship.

This is the traditional view. On the other hand archaeologists and modern textual critics say that the book of Exodus was not composed by Moses though it may contain material from his time. On this view the story of Moses, as we have it in Exodus, is a much later compilation. It is neither internally consistent, nor in agreement with the archaeological evidence, nor compatible with the records of contemporary people such as the Egyptians. Furthermore, there is the problem of the miracles which are to be found throughout the story. But there is no need for us now to explore these many difficulties, since the historical accuracy of the book of Exodus is largely irrelevant to our present concerns. It is not what happened but what was said to have happened, what was believed to have happened by the generations of Jews and Christians that we must understand. The very special notions of historical truth which have governed recent Biblical enquiry are themselves the products of a particular time and place, and are radically different from the notions of truth which have shaped the versions of the past which we have inherited from the past. It is with these earlier versions that we are concerned.

Moses moulded his people in the harsh conditions of the desert. As they frequently said themselves, even slavery in Egypt was to be preferred to the perils and deprivation which they encountered after it. Moses described their sufferings in Egypt as an iron furnace from which he had freed them, but that metaphor could stand for their time in the desert also. The leading of the people out of Egypt is nonetheless the supreme symbol in the Old Testament for the liberation of the spirit from the bondage of the body, to the extent that we still talk of the fleshpots of Egypt even though these people were fleeing slavery, not luxury. In this symbolism the crossing of the Red Sea marks the exact moment of transition from the bodily to the spiritual, a passing from one world to the other. As soon as the passage was accomplished the path was closed, and those who tried to follow were overwhelmed.

This crossing was also a crossing from the settled to the nomadic way of life, until the Jews should settle again in the land promised them. That they took forty years to reach that land was not part of the divine plan from the beginning but the result of their own errors. But from the very beginning of their journey beyond the Red Sea this whole people was spiritually exalted to a degree and for a length of time unparalleled elsewhere. Going into the desert for spiritual refreshment is a commonplace of the Semitic religions, but in this case an entire human population underwent this process of purification, more than a million people without previous experience of desert conditions. How frightening it must have been can be gathered from the people's lack of confidence in their leadership. Despite all the miracles, the parting of the sea, the manna, the events at Sinai, the quails, the rocks which gave water, the people were forever at the point of revolt. Moses himself had been given his mission in this same desert, at Mount Horeb and the burning bush, when he had fled Egypt after killing the overseer. But nothing could have prepared him for the task of leading an entire people to this place.

Exactly three months after leaving Egypt they came to Mount Sinai in the very midst of the wilderness. The reason for their going this way rather than along the coast was to avoid the armies of the Philistines. This choice of direction was providential, since the revelations made to Moses and his people at Mount Sinai were the source and foundation of Judaism. These revelations took place on a mountain, many of them in the sight of the whole people. The mountain connected heaven and earth, divine and human. This was not merely because the top of the mountain was close to the sky. Symbolically mountains represent the vertical line stretching from the highest to the lowest, along which all the dimensions of being are strung like beads on a thread.

It was on a mountain that Moses saw the burning bush and received from God the orders which began his mission. It was on top of a mountain that Noah's ark first made landfall after the flood, that ark which contained in principle all the creatures on earth. The mountain is an axis which represents in linear form the heart or centre where all things reside in principle. This is the place to which the craftworker must come in order to conceive the idea of the thing to be made, as Moses here received the commandments, the pattern of the tabernacle, and the patterns of all that it contained.

Visible from all sides, Mount Sinai served as an enormous stage on which the drama of the divine revelation could be seen by a whole people. Its visibility distinguishes the mountain from all other symbols of the sacred axis or centre. Because it is the most visible of all the things on earth, its use as a symbol of the centre corresponds to a time when the truth which it represents is open and available to everyone. This time is the first and earliest period of human wisdom, when 'all the Lord's people are prophets', before it becomes necessary to restrict access to the truth by imposing the requirement of initiation. When this occurs, the symbol of the sacred centre ceases to be the mountain and becomes instead the cave within the mountain, the chamber in the pyramid, the hidden place in which initiation occurs. Mount Sinai was at once easy and difficult of access, difficult because in the midst of a wilderness and easy being a mountain. Its isolation enabled the recreation of that time when the truth was accessible to all. Once the people reached Mount Sinai, they had already been purified and set apart from the human world around them.

Even so, the ordeal of the revelation was more than the people could bear. In a number of ways they fell short of what was required of them. Despite a further three days of ritual purifications and the setting of limits to how near they could come to the mountain, they were terrified by the thunder and lightning, the trumpet calls and the smoke which announced God's presence. God wanted the people to hear what he told Moses so that they would believe Moses for ever, but the first part of the revelation, the giving of the commandments by word of mouth, was more than the people could bear. They trembled and withdrew and asked Moses to tell them what God had said, because they were afraid that if they heard God directly they would die. Nonetheless there was a select group of seventy elders who went with Moses some way up the mountain and were there permitted a vision of God on the sapphire pavement.

God gave Moses the commandments and then prescribed the rituals and ritual objects which were to form the basis of Judaism. The commandments in some ways cut against the ritual prescriptions which followed them. For example, there was a strict ban on carved images, on any likeness of what was in heaven or on the earth or in the sea. But in the instructions for the sacred tabernacle, the mobile temple, God commanded that the gold lampstand was to be ornamented with almond blossoms. Perhaps the fact that these were hammered allowed them to escape the ban, but they were still likenesses. Or perhaps the point is that they were not worshipped for themselves but merely adorned the place where God spoke to his people. Another commandment forbade the making of gods out of silver or gold. Instead, any altar should be of earth or unhewn stone

for if you use your tool on it, you have profaned it.

This commandment is followed a few chapters later by God's description of the altars made from various metals which were to be placed in the tabernacle.

These contradictions may be explained in several ways. It may be that the Sinaitic revelation as we have it in Exodus is a conflation of traditions from different historical periods. Some scholars argue that the description of the furnishings and ornaments in the tabernacle does not belong here but is an anachronistic retrojection from the period of Solomon's temple or later. On this account we should suppose that the tabernacle was a much simpler arrangement, lighter, better fitted to the rigours of nomadic life. During this period of worship, the argument goes on, the metals were largely taboo for ritual purposes as the commandments suggest. We must remember, for example, that Moses was the first of whom it is explicitly said that he was circumcised by flint, and that this became the general practice at the time of Joshua, his immediate successor.

Or we may take it that the commandments and the prescriptions for the tabernacle do belong together but in an unusual way. The point of those commandments which appear to contradict what follows them is that they emphasise how the forms and furnishings of the tabernacle are unique. These commandments create, as it were, the space within which the prescriptions for the tabernacle gain their exclusive force. All graven images were forbidden except for those which were prescribed explicitly and in the greatest detail. This is a rather more likely resolution of the contradictions between the commandments and the prescriptions for the tabernacle than that they must have come from different historical periods. For then we must suppose that many generations of Jews and Christians have winked at an impossible anachronism, at a tabernacle made to contain commandments which it flagrantly breached. And it adequately explains God's wrath with the Jews for making a golden calf since this had not been authorised, while some of the altars which God directed Moses to make were horned.

First of all, God ordered Moses to make the ark of the covenant. The ark of the convenant was a wooden chest covered inside and out with pure gold. It was placed in the holy of holies, which was a perfect cube separated from the rest of the tabernacle by a curtain. In the ark were placed the two tablets of stone on which God had written the commandments. Four gold rings at the corners of the ark enabled it to be carried by wooden poles covered with gold, and these poles were never removed from the rings, so that they were as sacrosanct as the ark itself. The rings were made of gold, not pure gold. If they had been pure gold they would have been too soft for their purpose. There were other items in the tabernacle carried in this way, but only of the ark was it specified that the poles should never be taken from the rings nor touched by unconsecrated hands. This demonstrated the supreme holiness of the ark as the container of the commandments. The ark was overlaid both inside and out and in this too it was unique since all the other overlaid items in the tabernacle were overlaid on the outside only. And it was overlaid in this way with pure gold which again signified its special holiness, since the outer parts of the tabernacle were fitted with bronze or silver.

There were clear gradations in the design of the tabernacle, from bronze to silver to gold, corresponding to the gradations of holiness from the outer parts to the holy of holies. These same gradations corresponded also to the restrictions placed on access to the various parts of the tabernacle, from the outer court which was accessible to the laity, to the holy place to which only the priests and elders were admitted, and finally to the holy of holies which was reserved for Moses or the high priest. These gradations, symbolised by the metals, were also represented by the restrictions placed on access to Mount Sinai itself. In these ways a hierarchy was created at the same time as a whole people was sanctified.

The lid of the ark was made of pure and solid gold and above the ark God directed that there be made a mercy seat which was to be his throne. This had the same length and breadth as the ark and on each of the two ends was a cherub made of pure hammered gold, facing inward so that the two cherubim faced each other.

> And the cherubim shall stretch out their wings above, covering the mercy seat with their wings.

God told Moses that he would meet and speak with him from above the mercy seat, from between the two cherubim. In this way the ark itself became the base of the throne, in accordance with the ancient custom of placing a copy of a law or treaty at the feet of the god who ratified it. The cherubim were a constant feature of God enthroned. Their most important quality was that they were winged, though the number of their wings differed in different scriptures. They are usually taken to signify the mobility or omnipresence of God, a necessary corrective to the danger of specifying God's location in a single place. In order to overcome the temptation to restrict God to the holy of holies, the only ornaments in the holy of holies were a symbol of his being everywhere.

Besides the cherubim and the mercy seat there was one other large item of solid gold in the tabernacle. This was the golden lampstand or menorah in the holy place. God directed that this should be made from a talent of pure gold, and that all its parts and ornaments should be made from the one piece. This meant that its base, its shaft, its branches, bowls, ornamental knobs and flowers were all to be hammered out of a single ingot of gold. They were not to be made separately and then fitted together. The reason for this may have been that the lampstand was to symbolise in its manufacture the unity of the divine nature. Similarly God laid great stress on the importance of combining the different elements of the tabernacle so that the completed edifice became one whole. These elements were necessarily made separately and finally were brought together by Moses to create the unitary tabernacle. But in the case of the lampstand it appears that it was unitary from beginning to end. God also commanded that there be made wick trimmers and trays from the same ingot. These were separate from the lampstand and presumably enabled the maker or makers of the lampstand to cut away metal as well as to hammer it.

Nevertheless the manufacture of the lampstand must have been a formidable task if it was made in this way. When we add the extremely complex prescriptions for its branches, bowls, flowers and knobs, the difficulties begin to appear insuperable. There were a number of traditions which suggested just that, that the task could not have been accomplished quite as it is described in scripture. Some say that Moses found it impossible to remember the details of the lampstand and had to return several times to the subject, until finally God decided to instruct Bezaleel, the master craftsman, directly. According to others the lampstand was not a product of human craft at all. They base their theory on a passage in which God says that the lampstand will be made, rather than that Moses will make it. On this basis they argue that the lampstand was created simply by throwing the gold ingot into the furnace. The ingot then turned into the lampstand all of itself! This comes from the oldest Rabbinical commentary, despite the fact that it contradicts a later passage in Exodus which describes the making of the lampstand by Bezaleel. Still another theory is that the lampstand was cast and that the hammering was the finishing touch.

But whether it was made by Bezaleel or by that hand which wrote the commandments on the tablets of stone, this lampstand was the supreme metallurgical achievement of the Jewish tradition as the shield of Achilles was of the Greek. The question that remains is whether the lampstand like the shield was a symbol of the divine creation. It was certainly regarded as such in later periods, but this is one of those places where it may be necessary to distinguish between Jewish and Greek ways of thought. This point may be illustrated by the two quite different interpretations of the lampstand given by Josephus in the first century AD. In his account of the Jewish wars, written for the Jews when he was quite young, Josephus described the lampstand carried in the Roman triumph. He wrote here that its seven lights, the one on the central shaft and the three on each side, were an emblem of the dignity accorded to the number seven by the Jews. This is usually taken to mean that the seven lights of the lampstand represented the six days of creation and the first sabbath.

But in his later work on the antiquities of the Jews, written after he had been in much closer contact with gentile literature, Josephus interpreted the seven lights of the lampstand as representing the seven planets, the light of the central shaft being the sun. In this he agreed with his near contemporary Philo, the great Jewish scholar who devoted much of his life to showing how Greek and Jewish doctrines were finally the same. Whether the lights of the lampstand represented the seven days or the seven planets would not by itself make very much difference. Either way they symbolised the divine creation. If they represented the seven days we would have another example of the Jewish emphasis on sacred time rather than on sacred space, an emphasis peculiar to nomadic peoples. If on the other hand the lights represented the planets, then the sacred metallurgy of the Jews would be much more like that of the Egyptians and the Greeks. Or it may be that the seven lamps signified both the planets and the days of the first week, just as we still call the seventh day Sunday.

But these different ways of interpreting the lampstand apply to very much more than merely the lampstand. They can be made to apply to the entire furnishing and construction of the tabernacle and to the robes and ornaments of the high priest as well. Josephus, for example, after finishing his account of the lampstand in his work on Jewish antiquities, goes on to interpret many other features of the tabernacle as having a planetary or stellar significance. Since many of these features, like the lampstand, have to do with metals or precious stones, we need to understand their symbolic meanings since these are vital to an understanding of the spiritual significance of the metals in western thought. At this stage the most we can say is that from the time of Philo and Josephus there were Jewish schools of thought which interpreted details of the tabernacle astronomically. But there were other Jewish schools which did not, but interpreted these details in accordance with Jewish traditions only.

Though made of pure gold, the lampstand was not placed in the holy of holies but in the holy place just beyond the veil or curtain which covered the entrance to the holy of holies. With the lampstand in the holy place were an altar for incense and the table for the shewbread. Both of these were overlaid with gold on the outside only, unlike the ark which was overlaid within and without. In his later writings Josephus suggested that the altar of incense and the offerings made upon it symbolised the four elements of fire, air, earth and water. As for the twelve loaves of shewbread replaced each week, they represented for Josephus the twelve months or the twelve signs of the zodiac. For other Jewish schools they signified merely the twelve tribes of Israel. Beyond the holy place, in the outer court accessible to the laity, were a bronze altar for animal sacrifice, and a bronze laver in which the priests were to wash before they entered the holy place. Likewise the sockets of the outermost posts of the tabernacle were of bronze and so were the clasps which joined its outermost coverings. The interior sockets and clasps were of silver or gold.

After describing the construction and most of the furnishings of the tabernacle God told Moses how the high priest should be robed. This ceremonial dress made use of gold and precious stones. Its most impressive feature was the gold breast plate inlaid with twelve precious and semiprecious stones. These were arranged in four rows of three and inscribed with the names of the twelve tribes of Israel. There is an old tradition that these stones were not engraved with the names of the tribes but that they had been written thereon by a worm, the shamir. This story may be connected with the prohibition on hewing the stone of the altar and the later Solomonic commandment that no hammer or chisel or any iron tool be heard in the temple while it was being built. According to Josephus the stones of the breast plate shone with a special radiance when the Jews were about to win victory in battle. For this reason those Greeks who venerated the Jewish laws came to call it "the oracle". But Josephus added that this oracular power had been lost some two centuries before his own time as a result of Jewish apostasy. Nonetheless, if this tradition is true, the breast plate like the mercy seat allowed direct communication between God and his people. Moses at certain times was so radiant as a result of his communion with God that he had to wear a veil in public.

The breast plate and its stones were worn by the high priest when he entered the holy of holies. In this way they ritually introduced the twelve tribes into the presence of God. By means of the breast plate the whole people was symbolically enabled to enter that place reserved for Moses or the high priest alone. But at least from the time of Philo and Josephus, the twelve stones of the breast plate were interpreted as representing the twelve signs of the zodiac, in this way complementing the planetary symbolism of the lampstand. Enormous ingenuity and effort have been expended over the centuries in marrying the twelve tribes, the twelve stones and the twelve signs of the zodiac, since Exodus gives us only the names of the stones and their places in the gold base. Even the names of the stones are far from easy to interpret. Controversy over their identity has continued for millennia despite the fact that Josephus claimed to have seen the actual breast plate on many occasions. In some medieval and renaissance pictures and sculptures of the breast plate, the signs of the zodiac have replaced the names of the tribes on the stones.

On the headdress of the high priest was a plate of gold on which was written in the ancient Samaritan writing 'Holiness to the Lord'. This item at least was not interpreted astronomically. This plate for the forehead was preserved well into the first millennium of our era. Fringing the bottom of the long garment which reached almost to the high priest's feet were pomegranates and bells, the pomegranates made of yarn and the bells of gold. These alternated right round the bottom of the garment. The bells were explicitly described by God as a warning device so that those who heard them should know that the high priest was approaching. Such warnings were necessary because the tabernacle, like Mount Sinai, could prove fatal to those not properly prepared. Two of Aaron's sons, Moses' nephews, were destroyed there by fire for performing an unauthorised ritual. According to the later Josephus the bells on the high priest's robe were symbols of thunder, while the robe itself was a symbol of the sky.

The instructions for the making of the tabernacle, its vessels and the garments of the high priest were given to Moses during the forty days and nights he spent in the cloud and fire which covered the top of Mount Sinai. But Moses was not the only one of his people to be inspired directly by God to make these things. Moses was told that among the people there were two artisans, Bezaleel from the tribe of Judah and Aholiab from the tribe of Dan, who had been filled

> with the spirit of God, in wisdom, in understanding, in knowledge and in all manner of workmanship, to design artistic works, to work in gold, in silver, in bronze, in cutting jewels for setting, in carving wood, and to work in all manner of workmanship.

Furthermore, God told Moses that he had put wisdom in the hearts of all who were gifted artisans so that they might make everything he had prescribed.

The expertise for the manufacture of all these things was found among the people. Everything that had been prescribed could be made by the children of Israel themselves, though it was very different work from the making of bricks for which they had been used in Egypt. All those capable of contributing to the manufacture of the tabernacle were inspired to do so, so that to this extent the whole people was involved in the sacred work. This is another aspect of that sanctification of the whole people which was symbolised by the mountain and which was achieved by the public nature of the revelation. In the same way all the people were asked to contribute the materials for the making of the tabernacle and its ornaments, out of the store which they had taken from the houses of the Egyptians in the last moments of their captivity. They gave so freely that the artisans had to tell Moses to tell the people to stop giving, since they quickly had more than enough. And we must note that the wisdom to manufacture these things, which God gave to Bezaleel and Aholiab, was called the spirit of God. This is equivalent to the Greek and Roman theory of the divine origin of the crafts, which was typified in the stories of Prometheus and Arachne. Not only the ideas of the things to be made but the very means of making them were directly the work of the divine. This is the same spirit of God which moved on the face of the waters in the second verse of the Bible.

There is a close connection between the divine instructions for the tabernacle and the account of the creation in the first chapter of Genesis. The instructions for the tabernacle were in six parts, each beginning with the words 'The Lord spoke to Moses, saying...' At the end of these six sections God stressed the importance of keeping the sabbath. Once again we seen how the Jewish cosmogony was based upon sacred time, not space. In this light the planetary and stellar interpretations of Josephus and Philo seem less plausible. The lampstand is more readily understood to symbolise the days of creation when the entire tabernacle, as God described it, did the same. It is remarkable too that the tabernacle was finally erected on the first day of the first month. This was the day on which the waters of the flood began to recede from the earth in the story of Noah. And Moses consecrated the tabernacle as soon as it was set up, just as God blessed the world he had created when he saw that it was good.

SOLOMON AND DANIEL

Four hundred and eighty years after God directed Moses how to make and equip the tabernacle, King Solomon began work on the first temple of the Jews in Jerusalem. Though this temple was much more elaborate and costly than the tabernacle which it replaced, it lacked much of the spiritual and popular power of the tabernacle in its origin and in the manner of its construction. The site for the temple had been given to Solomon's father,

King David, who had also set aside gold and silver and precious vessels for it. But God had determined that David's son, not David himself, should build it. In essence the temple was the tabernacle writ large and established in a single place. But still the ark of the covenant was placed in the temple's holy of holies with its carrying poles in its rings, and with the poles clearly visible from the holy place beyond the curtain. The nomadic origins of Jewish worship were deliberately kept to the fore in the new, settled dispensation.

The temple was not designed by God as the tabernacle had been, but by Solomon himself. This is the single greatest difference between the temple and the tabernacle. When the temple was consecrated God demonstrated his presence in his temple in much the same way as he had in the tabernacle. The temple was covered with a cloud. But the various additions and alterations which Solomon introduced do not seem to have required divine authorisation from the beginning. This is significant when we remember that innovations in the ritual practice of the tabernacle could bring about the death of those who attempted them, as had happened to two of Aaron's sons. Unlike the arrangements for the tabernacle, the materials and labour for the building of the temple were not provided only by the Jews. The cedar for the temple came from Lebanon and the labour was forced from the one hundred and fifty thousand aliens resident in the land of Israel. This was very different from the labour and materials so freely given by the people that Moses had to stop their giving.

Perhaps strangest of all was the source of the expertise needed for the temple. In the case of the tabernacle that expertise had come from among the Israelites. Two of them, Bezaleel and Aholiab, had been filled with the spirit of God which gave them all the understanding they needed to make and to oversee the making of the different artifacts. These men were found almost immediately after the departure of the Israelites from Egypt where they had been employed as slaves in the making of bricks. But for the temple Solomon found it necessary to import a master craftsman, Hiram of Tyre. Hiram had a Jewish mother but he was firmly identified with Tyre and was sent from there to Solomon by the King of Tyre. And this happened at a time when the Jews were at the very peak of their economic and commercial success, when in Jerusalem King Solomon made silver as common as pebbles and cedars as plentiful as sycamores.

Many of the fine distinctions between the different parts of the tabernacle were not carried over into the building of the temple. In the construction of the tabernacle different metals had been used to distinguish the different degrees of holiness from the holy of holies to the outer court. Though the temple preserved the basic ground plan of the tabernacle, its inside was overlaid throughout with pure gold. In this way the symbology of the metals was partly discarded. The new holy of holies like the old was a perfect cube but several times larger. New cherubim were made whose wings now stretched to either side of the room and met in the middle, so that not only the ark but the entire room was covered by them. When the temple was finished, the ark and all the old utensils were ritually carried into it in procession and took their place alongside the new creations of Solomon. Solomon seems to have made new and larger versions of almost everything in the tabernacle except the ark itself. There were larger altars with all their bowls and instruments, a larger laver, ten golden lampstands 'made in the way prescribed', and at least one novelty, two pillars of bronze festooned with pomegranates, to stand in the holy place.

Nearly four hundred years later these same vessels of the temple were taken as spoil to Babylon, according to the first verse of the first chapter of the book of Daniel. Nebuchadnezzar, prince of Babylon, laid siege to Jerusalem, reduced it and then took the temple vessels to the treasure house of his god in Babylon. The transportation of all the valuables required several journeys. At the same time he took certain royal young men from Israel to teach them the learning and language of the Chaldeans. Among these young men was Daniel. As far as we know, the temple vessels remained in the Babylonian treasury throughout Nebuchadnezzar's reign and for some time after it. But in the fifth chapter of Daniel we hear of these vessels again. By this time Belshazzar was on the throne of Babylon and Daniel was about ninety years old. Belshazzar made a great feast for a thousand of his lords and ordered that the golden vessels which Nebuchadnezzar had taken from the temple in Jerusalem should be used by the feasters. This was done and Belshazzar with his princes, wives and concubines profaned the sacred vessels. They drank wine from them and praised their own gods of metal, wood and stone as they did so.

This act explicitly defied the God of the Jews, but Belshazzar's reasons for doing this are not given. Perhaps he had been told that the Jews believed his kingdom would be overthrown. Whatever his reasons, the consequences of his act were immediate. In the same hour, we are told, the fingers of a man's hand appeared, and over by the lampstand they wrote some words on the plaster of the wall. When the king saw these fingers and what they wrote, his expression changed and his knees began to knock. He cried out for the wise men of the Chaldeans to be brought in, to explain what the fingers had written. The words on the wall could be seen by everyone but we are not told how many saw the fingers which wrote them. Belshazzar did, and they wrote over against the lampstand for maximum visibility. And perhaps they wrote there for another reason. This lampstand is mentioned at a time when the vessels from the temple of Jerusalem were being profaned at a banquet. The lampstand may well have been one of those vessels, either one of the ten lampstands of Solomon, or even the original lampstand of Moses and Bezaleel. It was near the king's table, in a place of honour. In its light the fingers wrote on the wall, fingers which came from God like those which wrote the commandments on Mount Sinai. That writing was kept in the ark and the ark was housed in the temple or the tabernacle near the golden lampstand just beyond the curtain.

The wise men whom Belshazzar summoned to read the words written on the wall proved unable to do so, even when a third of the kingdom was offered to the one who should succeed. At this Belshazzar and his thousand lords were even more distraught. At length they remembered Daniel and his previous success as an interpreter of dreams during the reign of Nebuchadnezzar. An old man of ninety, Daniel was brought in and offered a third of the kingdom if he should succeed where the wise men had failed. He began by refusing this offer and then reminded Belshazzar of Nebuchadnezzar's downfall, who for seven years was like a beast, eating grass and living with the wild asses. God had done this to Nebuchadnezzar to humble him, and Nebuchadnezzar had indeed been humbled. But now, Daniel continued, Belshazzar had lifted himself up against the Lord of heaven and had profaned the sacred vessels in honour of false gods. So the true God had sent the fingers to write on the wall. Then Daniel interpreted the writing to mean the imminent end of Belshazzar's kingdom. Belshazzar clothed Daniel in scarlet and gold and made him ruler over a third of the kingdom. But that very night Belshazzar was killed and the kingdom of Babylon fell to the Medes. We know from elsewhere that Cyrus and Darius, the rulers of the new empire, ordered the return of all the sacred vessels from Babylon to Jerusalem and the rebuilding of the Jewish temple.

Daniel had acquired his reputation as an interpreter of dreams early in the reign of Nebuchadnezzar, to whose court he had been taken after the fall of Jerusalem. In the second year of his reign Nebuchadnezzar had a dream which troubled him. He summoned the magicians, sorcerers, astrologers and Chaldeans and asked them what he had dreamt as well as what it meant. The asking of the first question as well as the second put them in difficulties, but Nebuchadnezzar insisted that he had forgotten his dream and threatened to cut them in pieces if they did not answer. When they remained silent the captain of the guard was told to kill them, among whom Daniel and his companions, as students of Chaldean learning, were included. Daniel and his companions asked for more time and prayed to God, and Daniel had a night vision. The next day Daniel presented himself to Nebuchadnezzar and told him his dream.

> You, O king, were watching; and behold, a great image! This great image, whose splendour was excellent, stood before you; and its form was awesome.

This image's head was of fine gold, its chest and arms of silver, its belly and thighs of bronze.

Its legs of iron, its feet partly of iron and partly of clay.

You watched while a stone was cut out without hands, which struck the image on its feet of iron and clay, and broke them in pieces.

Then the iron, the clay, the bronze, the silver, and the gold were crushed together, and became like chaff from the summer threshing floors, the wind carried them away so that no trace of them was found. And the stone that struck the image became a great mountain and filled the whole earth.

Having given his account of Nebuchadnezzar's dream Daniel then proceeded to its interpretation. According to Daniel the golden head of the image represented the kingdom of Nebuchadnezzar himself. The silver and bronze parts of the image represented later kingdoms, inferior to that of Nebuchadnezzar but still ruling over the whole earth. They in turn were to be succeeded by a fourth kingdom as strong as iron. But this kingdom, though it appeared strong, had great weaknesses, as indicated by the clay. In the time of this kingdom God would set up an indestructible kingdom of his own which would stand forever. This was represented by the stone cut out from the mountain without hands, which brought the whole image crashing down. When Nebuchadnezzar had listened to all this, he fell upon his face and worshipped Daniel, made him ruler of the province of Babylon and the chief of his wise men. He had recognised his dream just as Daniel described it.

Why was the history of these empires realised in the figure of a man? What exactly was the significance of the different metals, the clay and the stone? What were the five empires described in Daniel's interpretation of the dream? There have been several different answers given to this last question, of which the oldest and best attested is that the silver parts represented the empire of the Medes and Persians, the bronze parts that of the Greeks and Macedonians, while the iron represented the Roman empire. Then the stone which brings the statue crashing down and grows to the size of a mountain is Christ. This is a very plausible answer but it carries with it the consequence that the dream and Daniel's interpretation of it were truly prophetic. For, however late we date the book of Daniel, there can be no question but that it was composed before the time of Christ, indeed before the consolidation of the Roman empire. Alternatively, the kingdoms may be identified differently, by counting either the Medes and Persians or the Greeks and Macedonians as two empires, not one. In this case the kingdom of iron can be dated to the second century BC, the time when the book of Daniel is supposed by many to have been written. Then the stone becomes simply another unfulfilled Messianic prophecy.

But, however we identify the empires, the fact remains that they correspond to the metals and to the parts of the human body. The obvious implication of identifying the golden head with Nebuchadnezzar's kingdom is that this kingdom was superior to the ones which succeeded it as the head is the highest and best part of the body, and gold is the best of the metals. This implication was made explicit in Daniel's interpretation when he said that the kingdom after Nebuchadnezzar's, the silver kingdom, would be inferior to the one before it. So we may infer that the four metals, gold, silver, bronze and iron, were ranked in order of excellence, and that the clay was inferior to all four as indicated by its being part of the feet. The only difficulty with this inference is the stone which destroyed the image. For if the kingdoms are to be judged according to the quality of the materials which represent them, then the kingdom represented by stone should be inferior to those represented by the metals, since stone even more than clay is unworked and unrefined. The stone's being unworked in any normal way is emphasised in the dream where we are told that it was cut out without hands.

It is possible to read the Hebrew words which describe the stone's separation from the mountain as meaning no more than that the stone comes away from the mountain without human intervention, that it merely rolls down the mountainside for example. That the kingdom of the Messiah should be represented by a stone which brings about the destruction of kingdoms represented by metals, and that the stone should move of itself, are ways of showing the differences between the kingdom of the Messiah and worldly kingdoms. We remember in this context that this king was crowned with thorns, and was not enthroned but crucified. There are other passages in the Bible where Christ is spoken of as a stone, as the stone which the builders rejected but which has now become the corner stone of the temple. This cornerstone, however, must have been hewn and is so spoken of in some of the texts. But the stone which destroyed the great metal figure was not hewn by human hands and this recalls the commandment given to Moses in the book of Exodus, that the stone of the altar should not be worked by a tool lest it be profaned. In the tension between this commandment and the prescriptions for the metal altars which follow it, we have perhaps a precursor to the tension between the stone and the metal figure which it destroyed in the book of Daniel.

The metal figure is equally problematic and has inspired an enormous body of commentary. Fortunately we are not here concerned with the identification of the different empires but with the symbolism of the metals. The most noteworthy feature of these metals is that it is very hard to see why they should have been selected and ordered in this way. Gold and silver are perhaps to be expected though it is by no means always the case that gold is more valuable than silver. Silver is, however, tarnishable and is inferior to gold in this respect. But why bronze, which is not even a pure metal but an alloy of copper and tin? We may say, of course, that the bronze and iron of the figure indicate different stages of technical development. But then there is the problem that gold and silver cannot be taken to represent stages of technical development in this same way. The simple fact is that the golden age, as the ancients thought of it, does not belong to the same order of understanding as the bronze or iron ages as we think of them. Nonetheless the ancients did suppose that certain periods of time corresponded to bronze and iron as other periods corresponded to gold and silver. And we have already seen how the construction of the tabernacle made use of gold, silver and bronze to signify different degrees of holiness, while iron seems to have been excluded from Jewish religious use as far as possible.

Whatever reasons the Jews may have had for singling out these metals and arranging them in this order, they were not the only people to do so. In the seventh century BC the Greek poet Hesiod used these same four metals in the same order to represent or characterise four different human races made by Zeus, the king of the gods. The different races here suggest a much vaster scale of time than do the empires described by Daniel, but the similarities between the two stories are striking. In both stories the same metals are ordered in the same way to indicate a series of declines in quality. In the Hesiodic story there is no continuity between the different races, but each requires a separate creation; in Daniel's figure the different metals pass into one another so as to make the one statue. This gives the sense of an historical continuum and emphasises again the sharp contrast between the metal empires in their relations to each other and the stone empire in its relation to them all. As we have seen, the bronze and iron empires of Daniel may have signified empires which differed from each other in their use of these two metals. In Hesiod's story this is made quite explicit: the bronze race of humans had no black iron at all.

The author of the book of Daniel may have derived his ideas about these metals and historical periods from the Greeks, especially if we date the book as late as the second century BC as is often done. Or the Greek and Jewish theory of these metals may derive from some other source common to both. But it is, I think, reasonable to suppose some connection between the thinking of these two peoples on this point since the choice and ordering of these metals to represent the passage of long periods of time is by no means obvious. Plato also made a contribution to this way of thinking. He divided society into four classes corresponding to these same four metals and then interpreted the process of an historical cycle as the devolution of power from the highest class to the lowest. Plato did not identify his four social classes with the parts of the human body but only with the metals. Other traditions such as the Indian also divide society into four classes, classes very like Plato's, and then identify these classes with the parts of the body. If for a moment we conflate the Platonic and Indian theories of the social classes we arrive at the following table:

Metal	Class	Part of Body	Polity
Gold	Philosophers Priests	Head	Theocracy
Silver	Nobles Warriors	Arms, Chest	Aristocracy
Bronze	Farmers, Artisans Merchants	Belly	Plutocracy
Iron	Labourers	Legs	Democracy

Conflations of this kind are of dubious value, and there are important differences between the Platonic social classes and those of the Indians. Nonetheless, this table does demonstrate a way of identifying the parts of the body, the four metals and the four stages of an historical continuum, all of which we are given in Daniel's account of Nebuchadnezzar's dream and the interpretation of it. To be sure, we do not find in the book of Daniel any reference to social classes, but only to empires. Yet it is hard to resist the feeling that this passage from the book of Daniel belongs in the context of a much more general theory of correspondences between the metals, the parts of the body, societies and historical change. At the very least it is easier to see in the light of Hesiod and Plato how Nebuchadnezzar could have come to dream of a single human figure made of different metals, when really he was being given an insight into the rise and fall of empires.

The dream of Nebuchadnezzar is one of the most impressive examples of metallic imagery in the Bible, but it is only one of the metallic symbols in the book of Daniel. And in the story of the fiery furnace we are given a metallurgical and not merely a metallic symbol. The story goes that three companions of Daniel refused to bow down to a golden image set up by King Nebuchadnezzar. They were informed upon by certain Chaldeans who were envious of the promotions given to Daniel and his friends for his having solved the problem of Nebuchadnezzar's dream. The king orders the three to be thrown into a fiery furnace, heated seven times hotter than usual for the occasion. Those who cast the three into the flames are killed by the heat. But Daniel's companions are unaffected. They walk about in the midst of the burning furnace and Nebuchadnezzar sees them there, and with them a fourth figure. He calls them out and the three emerge completely unchanged by the fire. Then Nebuchadnezzar relents and promotes them still higher in Babylon.

The testing and refining of the soul through suffering as in the fire of a furnace is a very common metaphor in the Old Testament, from the first five books to the minor prophets. With the wine press, the furnace is one of the great symbols of tribulation. As the book of Proverbs puts it:

> A crucible for silver, a furnace for gold, but God for the testing of hearts.

In the case of Daniel's companions, an actual furnace which had been heated several times hotter than usual was powerless to affect them. The reason for their incombustibility is made quite clear. They had refused to worship Nebuchadnezzar's image, and when threatened with the furnace by the furious king himself, had calmly refused again on the grounds that their God would protect them in the furnace if he so wished. If not, they would still not worship the image. They think it possible that God will save them, but they do not necessarily expect him to. It is this complete obedience to divine providence, even to the point of death, which paradoxically saves them. To put it in the language of metals, they are already completely refined and purified; their obedience to the divine will is absolute; there is no dross in them and therefore the heat of the furnace is powerless to affect them. As the Chinese say, 'Real gold fears not the fiery furnace.'

Who was the fourth man whom Nebuchadnezzar saw walking with Daniel's companions in the midst of the furnace? At first the king described him as being like the son of God, and later he described him as an angel. The early fathers of the Church supposed him to be Christ himself. There is certainly a resemblance between this passage and the one describing the disciples' journey to Emmaus after the crucifixion, when they are joined by another traveller whom they do not recognise but who later reveals himself to them as Jesus. Before his death Jesus told his disciples that where two or three were gathered in his name he would be among them. Other commentators identified the fourth figure as the angel of the presence whom God appointed to look after the Jews in the desert. Later in the book of Daniel the angels are identified by name, as for example Gabriel, but there is one figure in Daniel's visions who is not so identified. This figure is much more fully described than the angels, he directs the angels to do his bidding and the sight of him fills Daniel with terror. He too is often thought to be the Christ or the Logos. This is how he is described:

I lifted my eyes and looked, and behold a certain man clothed in
linen, whose waist was girded with gold of Uphaz!
His body was like beryl, his face like the appearance of lightning,
his eyes like torches of fire, his arms and feet like burnished bronze
in colour, and the sound of his words like the voice of a multitude.

It certainly sounds as though he might be the figure in the furnace.
As we shall see, this visionary figure reappears in the last book of the
Bible.

THE NEW TESTAMENT

When Jesus died, the curtain between the holy of holies and the holy
place was torn in two from top to bottom. This signified the end of the old
covenant between God and Moses, the covenant of the commandments,
and the beginning of a new covenant mediated by Christ. The word 'Christ'
meant 'the anointed one', one on whom the holy chrism had been poured
and who had been consecrated with blood as Aaron, the first high priest,
had been consecrated by Moses. By virtue of this consecration Aaron became the representative of the people in the presence of God. Into God's
presence he wore the twelve stones of the breast plate which symbolised
the twelve tribes. Clothed in ritual garments he passed through the curtain
into the holy of holies and returned, carrying God's commands as spoken
from above the mercy seat between the cherubim. The rending of the veil
signified the ending of this priesthood and its replacement by a new priesthood of Jesus, who was not consecrated with the blood of animals but with
his own blood on the cross. That the holy of holies was fitted with a curtain and not a door seems to have prefigured the moment when it would
cease to be the sacred centre. Like the ark, the curtain was covered with
cherubim who symbolised the omnipresence of God at the same time as
they decorated his special place in the temple or tabernacle.

The sprinkling of blood on the covenant, the people and the high priest,
together with the consecration of the tabernacle and the temple, brought
about God's entry into the holy of holies in the times of Moses and Solomon. The holy of holies was the sacred centre of the tabernacle or temple
which were themselves the sacred centres of the Jews in the desert and in
the promised land. In the book of Numbers we are told how the tribes of
Israel pitched their tents in formation around the tabernacle as their centre.
The divine presence was indicated by a cloud which invested the tabernacle
and the temple immediately after they were consecrated. The sacredness of
the holy of holies in the tabernacle was, as we have seen, carefully stressed
by the symbolic use of metals. But with the death of Jesus this sacred
centre was outgrown. The rending of the curtain ended the period during
which God was localised in a particular place and symbolised by inanimate

objects. His presence on earth was now realised in the person of Jesus who had been consecrated with his own blood as he was being crucified. Jesus was not consecrated by a man, as when Moses consecrated Aaron. His consecration was directly authorised by God and was brought about by the passion. The very moment of that death was also a birth, as the divine presence burst out from within the holy of holies and a new spiritual order began. That this death was also a birth is indicated by the other detail we are given of the moment when Jesus died. Immediately after the rending of the curtain, the earth quaked and graves opened and many bodies of the saints who had fallen asleep were raised.

At the moment of Jesus' death all the fittings of the temple and the temple itself were made redundant. Paradoxically, of course, it is only by reference to these now redundant symbols that we can understand exactly what Jesus' death meant. We understand the rending of the veil only when we understand the significance of the holy of holies. In this way a knowledge of the book of Exodus is vital to the understanding of Christ's passion. The new order and the old were inextricably bound together, since the Christian story realised the hopes and anticipations of Judaism. This is the grand argument of the Epistle to the Hebrews, the authorship of which was traditionally attributed to St Paul.

The Epistle to the Hebrews puts the matter as follows:

> For when Moses had spoken every precept to all the people according to the law, he took the blood of calves and goats, with water, scarlet wool, and hyssop, and sprinkled both the book itself and all the people, saying, "This is the blood of the covenant which God has commanded you."
> Then likewise he sprinkled with blood both the tabernacle and all the vessels of the ministry.
> And according to the law almost all things are purged with blood, and without shedding of blood there is no remission.
> Therefore it was necessary that the copies of the things in the heavens should be purified with these, but the heavenly things themselves with better sacrifices than these.
> For Christ has not entered the holy places made with hands, which are copies of the true, but into heaven itself, now to appear in the presence of God for us.

The author of the epistle goes on to point out that unlike the high priest who had to perform the sacrifice of atonement every year, Christ had only to perform it once. Christ's sacrifice of himself took away the sins of many and came at the end of the ages. When he returned, it would not be to repeat the sacrifice.

The epistle refers explicitly to the vessels of the ministry and a little later mentions the copies of the things in the heavens. In this context it is clear that these copies are the vessels of the ministry just mentioned and perhaps the tabernacle itself. The phrase 'the copies of the things in the heavens' is consistent with the view of Philo and Josephus that the items of tabernacle and temple were astronomical symbols. That may be the force of the plural 'heavens'. As opposed to these copies, Christ entered heaven itself by his sacrifice, a singular 'heaven' which suggests the realm of the spirit. In this way the contrast between the Jewish ritual of purification and Christ's sacrifice is made as sharp as possible: animal blood against human, the blood of others against one's own, the realm of the heavens against the realm of the spirit, the copies of things against the things themselves, the holy places made by hands against what is really holy.

These powerful verses make explicit what the rending of the curtain symbolised. By sacrificing himself Christ made the forms of Judaic worship redundant. The laws, the ark, the temple and all its vessels were now dispensable. In their place was the figure of Christ on the cross. The elaborate vessels, garments and rituals, all of them prescribed directly by God to Moses on Mount Sinai, had completed their work of preparing the Jews for the Christian revelation, and they might now be discarded. Christ utterly transcended them at the same time as his transcendence was demonstrated in terms of them. No longer was there to be a holy of holies, nor an ark so sacred that even the poles by which it was carried could be touched only by consecrated hands. By the rending of the curtain the distinction between the holy of holies and the rest of the temple was abolished. This meant the end of those gradations of holiness which had been carefully instituted in the plans of the tabernacle and temple, in the series of restrictions on access to their various parts, and in the use of more valuable metals for their holier places. In this way the new Christian order dispensed with the sacred symbology of the metals which had preceded it.

The Christian revelation was absolute. Compared to the person of Christ nothing whatsoever was to be considered more holy than anything else. It is as though the distance between the crucified Christ and any other sacrifice or sacred object was so enormous that it made every other distinction between degrees of sacredness irrelevant. So far as holiness went, everything in the world was at the same distance from Christ as everything else, so that in this respect all distinctions between them disappeared. This point of view distinguished the Christian from every other philosophical and theological doctrine in the first centuries after Christ. Against the old and new pagan theologies of nature the Christian gospel proclaimed that through Christ's sacrifice the spirit had completely transcended all physical forms. The Church divested nature of all her gods, and the worship of nature and natural objects was declared sinful. Some historians of religion believe that it was precisely the clarity and force of this difference between the early Church and all the competing cults of its time which gave the Church its victory.

The Christian revelation was absolute in other ways. Just as Christ's sacrifice did away with the rituals and sacred objects of Judaism, so Christ's way of life displaced all the other ways of life in the society from which he came. To follow Christ required the surrender of everything, including family and work. It is sometimes said that Christianity is unique among the world's religions because its demands upon its followers are absolute in this way. In other religions there are prescriptions for many shades of belief and capacity, different rules and expectations for different degrees of commitment and ability. No doubt there is the assumption that obedience to one such set of prescriptions will prepare the devotee for a more arduous set, but there is always a recognition that there are many different ways of life and that almost all of them are worthy to some extent. Plato organised the different professions and occupations into four and then nine distinct categories, of which the highest, the philosophical life, would secure a final liberation from mortal existence if lived three times in succession. But for Plato the lives of the craftworker or soldier were still paths to the divine, imitations and manifestations of certain aspects of the divine at the human level. For the Greeks and Romans every craftworker cooperated to some degree in the work of Hephaestus or Athene or some other god, as Adam had cooperated with God in tending the garden of Eden.

But according to the Gospels there were no gradations or degrees in the service of Christ. Though Christ was brought up by a carpenter and showed himself to be a scholar of genius at a very early age, he renounced all this to prepare himself for his ministry. For him it seems to have been a matter of all or nothing. And this same life he enjoined upon his disciples, requiring them to leave everything and to live on what they were given. This requirement was somewhat tempered in the epistles where we are told that Christ's disciples should continue to practise their trades as a means of supporting themselves, but in the actual teaching of Christ there seems to have been no room for any other kind of life than that of complete surrender. His disciples were to be fishers of men, not fishermen. As a poet and teacher of the renunciate life there is no one to equal him, but this same transcendence which left the early Church without a doctrine or symbology of nature, left it also without a doctrine of work. These deficiencies, if that is what they were, were very quickly supplied, and the teaching of Christ was more or less accommodated to the needs of the social order. But the inescapable fact remained that neither work nor nature was much discussed by Christ himself, and this was to have profound consequences.

For this reason the New Testament hardly left room for a sacred theory of the metals or the metal crafts. In this it was very different from the theologies of the Greeks, Romans and Jews. The means by which the Church reinstated such a theory and then lost it again are the subjects of later chapters in this book. Nevertheless, even where the distinction between the old and the new covenants was made most sharply, there were important continuities between Jewish and Christian doctrine and forms of worship. The unhewn stone of the altar in Exodus and the stone cut out without hands in the book of Daniel prefigured the 'formlessness' of Christ and Christian worship. Both these stones were contrasted in their simplicity with the elaborateness of metal work. The commandment concerning the unhewn altar specified that no iron should be used on the stone lest it profane it. The stone in the book of Daniel damaged the feet of the great metal image which was then smashed, pulverised and blown away. In the same way Christ's sacrifice did away with the paraphernalia of Jewish worship.

Similarly both the Mosaic and Christian covenants emphasised their universality. The Mosaic covenant was not between God and one man nor one class of people, but with a whole nation. Though there were gradations of access to the mountain, and later to the tabernacle and temple, the revelation was made to everyone without any distinction between tribes or occupations. A whole people was inducted into these mysteries; all the Lord's people became prophets. The same was true of Christ's mission, which transformed the hierarchies of the Jewish priesthood and its distinctions between priest and laity. But in one respect Christ went further than the revelation on Mount Sinai, since that revelation inspired the Jewish craftworkers to the making of the tabernacle and its contents. There was at least one special dispensation of the spirit to enable this work to be done, and one class of people was chosen to stand in a special relationship with the divine. But in Christ's teaching neither the craftworker nor any other class was elevated in this way.

Despite the absolute nature of the New Testament in these respects, it yet began and ended with metal symbols. In the second chapter of St Matthew's gospel the new born Christ was given gifts by the wise men from the East, gold, frankincense and myrrh. Tradition has it that the gold symbolised Christ's kingship, the incense his divinity, and the myrrh his mortality. But though gold, the first of the three gifts, is certainly a symbol of royalty, it is also in the Jewish tradition a symbol of divinity, as it is in the tabernacle where pure gold was reserved for the most sacred objects. Likewise we may recall the gold of Havilah mentioned in connection with the garden of Eden in the second chapter of Genesis. This gold suggests the golden age of our first parents, which was to be renewed or restored by Christ, the second Adam, who would atone for the fall of the first Adam and show the way back to paradise. As the life of Christ began with a gift of gold, so it ended with iron or bronze, the nails through hands and feet which held him to the cross, and the head of the spear which wounded him in the side. Painters sometimes represent the swaddling clothes of the Christ child as the loin cloth of the figure on the cross. In something of the same way these different metals measure the distance between his birth and his death.

In St John's book of Revelation, the last book of the Bible, the figure who addressed Daniel in his visions and whose description I have already quoted, reappears. In St John's description the connection with the furnace is made explicit. His feet are like fine brass, as if refined in a furnace, and he stands in the midst of seven lampstands. The gorgeousness of this lighting recalls the description of the great palace by Homer, filled with metal work and lit at night by the torches held by golden statues. The figure in St John's vision is also explicitly Christ; he calls himself the first and the last, the one who lives and was dead, who is now alive for evermore. This figure dictates to St John the letters to be sent to the seven churches and conducts him to the second part of his vision.

In the penultimate chapter of Revelation, and of the whole Bible, St John sets out his vision of the New Jerusalem which is also called the tabernacle of God. He explains how an angel

> carried me away in the Spirit to a great and high mountain, and showed me the great city, the holy Jerusalem descending out of heaven from God, having the glory of God. And her light was like a most precious stone, like a jasper stone, clear as crystal.
> Also she had a great and high wall with twelve gates, and twelve angels at the gates, and names written on them, which are the names of the twelve tribes of the children of Israel:
> three gates on the East, three gates on the North, three gates on the South, and three gates on the West.
> Now the wall of the city had twelve foundations, and on them were the names of the twelve apostles of the Lamb.
> And he who talked with me had a gold reed to measure the city, its gates, and its wall.
>
> And the city is laid out as a square, and its length is as great as its breadth. And he measured the city with the reed: twelve thousand furlongs. Its length, breadth, and height are equal.
> Then he measured its wall: one hundred and forty four cubits, according to the measure of a man, that is, of an angel.
> And the construction of its wall was of jasper; and the city was pure gold, like clear glass.
> And the foundations of the wall of the city were adorned with all kinds of precious stones: the first foundation was jasper, the second sapphire, the third chalcedony, the fourth emerald,

the fifth sardonyx, the sixth sardius, the seventh chrysolite, the eighth beryl, the ninth topaz, the tenth chrysoprase, the eleventh jacinth, and the twelfth amethyst.

And the twelve gates were twelve pearls: each individual gate was of one pearl. And the street of the city was pure gold, like transparent glass.
But I saw no temple in it, for the Lord God Almighty and the Lamb are its temple.
And the city had no need of the sun or of the moon to shine in it, for the glory of God illuminated it, and the Lamb is its light.

If the vision of Christ among the seven lampstands recalls not only Daniel's vision but Homer's description of the palace at night, this account of the New Jerusalem has much in common with Plato's account of 'the real earth' which we considered at the end of the first chapter. Plato's afterworld was also replete with precious stones and metals, shining much more clearly than anything ever can on our earth. It may be, as I suggested, that the Platonic vision was a metaphor for the world of ideas, the inner realm of contemplation. In a similar way the New Jerusalem of St John descends from heaven and is illumined by its own light. The gold of which it is composed is like clear glass, transparent. Its receptivity to the light of the Lamb far exceeds the shining of things we know here, because it is seen in the spirit and not by the eye. But unlike Plato's vision, the New Jerusalem is a city and a tabernacle. Furthermore, its gold and precious stones, and the writing upon its gates of the names of the twelve tribes, remind us irresistibly of the high priest's breast plate. The stones are not quite the same ones, they are not themselves inscribed with the names of the tribes, and they are arranged in a different pattern, but for all that, the New Jerusalem is the breast plate writ large. In the midst of the city is the Lamb of Christ whose light illumines it. So this city is the most perfect synthesis of the Mosaic and Christian covenants, uniting in a single image the sacred objects of the tabernacle and the sacrifice of Christ.

But the tabernacle and the breast plate are less important to the understanding of this visionary Jerusalem than the garden of Eden which is its polar opposite. The Bible is stretched between the polarities of garden and city, vegetable and mineral, from the earthly paradise to the heavenly Jerusalem. The garden, we may take it, is circular, the cross section of a sphere against the cubic form of the visionary city. The cube is universally the symbolic form of the earth, the most stable of the elements. Likewise the city is the most settled form of human life, while the minerals and metals of the New Jerusalem, although they are sublimated and spiritual, are nonetheless representative of elemental fixity. It is the durability of these stones and metals which enables them to symbolise the unchanging and the eternal; their static quality compounds the squareness and stability of the city. In these ways the heavenly Jerusalem signifies the end of the wheel of time. There is no night there and neither sun nor moon.

This is the city at the end of time as the garden of Eden marked its beginning. In that garden a river sprang which divided into four rivers. These four rivers are four radii at right angles to each other, dividing the circular garden into four quadrants. According to the principles of sacred geometry, the square and the circle are inversions of each other. Turned inside out, the four quadrants of the circle makes the square. In this way the heavenly Jerusalem is the earthly paradise turned inside out. This is the symbolic meaning of that perennial problem, the squaring of the circle. The gold of Havilah, which one of those four Edenic rivers surrounded, is the gold of the heavenly Jerusalem. In the midst of this spiritual city is the tree of life, and the city is illumined by the Lamb, the new Adam, the sacrificial Christ.

Chapter Four

THE MEDIEVAL SYNTHESIS

> To me, I confess, one thing has always seemed preeminently fit-
> ting: that every costlier or costliest thing should serve, first and
> foremost, for the administration of the holy eucharist. If golden
> pouring vessels, golden vials, golden little mortars used to serve,
> by the word of God or the command of the prophet, to collect the
> blood of goats or calves or the red heifer: how much more must
> golden vessels, precious stones, and whatever is most valued among
> all created things, be laid out, with continual reverence and full
> devotion, for the reception of the blood of Christ.

Abbot Suger

IN THE FIFTEEN CENTURIES between the life of Christ and the Reformation, the sacred history of mining and metallurgy is difficult to trace. For we are no longer dealing, as we were in the first two chapters, with two distinct traditions, each of which was more or less self contained and each of which made constant reference to its own earlier phases. In the early Christian and medieval worlds we do not have a single tradition but a complex, in which the Greek and the Roman, the Jewish and New Testament elements are being woven and rewoven in a variety of designs. Nor are these the only elements. As the Christian world expanded and more and more of the European cultures were brought within its ambit, each contributed its own forms and fashions to the new international order. In many of these cultures the arts of working metal already had a long history, and were essential to their sacred traditions as they had been to the traditions of the Greeks, the Romans and the Jews. There is for example the magical hammer, Mjolnir, of the Norse god Thor. This hammer made Thor invincible, it had been forged by cunning dwarfs, and it returned to Thor's hand wherever he threw it. This hammer, represented by a T cross was the great symbol of Thor who was the most widely revered god in northern Europe before its conversion to Christianity. At that time the symbol of the hammer was simply adapted to represent the cross of Jesus. In such ways did the sacred traditions of prechristian Europe merge and reemerge in Christendom.

76

But within this dazzling array of traditions and cultures as they combined and recombined under the influence of Christendom, there were some continuous and universal problems. These problems arose out of the difficulty which we noted at the end of the last chapter, the difficulty of accommodating that absolute spirit of renunciation in Christ's teaching to the affairs of the social order. To be sure, this problem was in some ways less acute than it had been in Judaic monotheism. The invisible, ineffable, inconceivable God of Moses imposed an enormous strain upon his worshippers by his absolute unlikeness to them. How could those later times worship a divinity so far removed from anything within their experience? This strain upon the spiritual constitution of the Jews showed clearly enough even in the episode of the golden calf when Aaron, Moses' brother, yielded to the importunities of the people and gave them a visible, and therefore illicit, representation of Jehovah. And this people had directly witnessed God's power on many occasions. Exactly the same problem arose again and again in Christian practice, even though the new dispensation realised the divine nature in the person and actions of Christ. Once again there seemed to be an unbridgeable gulf between the purity and detachment of Christ's life and the workaday world of his followers.

As Christianity spread to the west this problem became even more acute, since the first western converts belonged in a traditional order which endowed almost every aspect of the natural and human worlds with divinity. The emphasis on the transcendence of God distinguished the Christian gospel from the major religions with which it competed in the first centuries of the new millennium. This difference between Christianity and its competitors may have been critical to its success. But in the longer term this transcendence and the renunciation of the world which it entailed proved impossible to maintain with the same rigour in later centuries. Slowly the older, prechristian ways of realising God in the world reemerged, but this reemergence was now tempered by the need to accommodate the old ways to the transcendence of the Christian revelation.

For the Greeks and Romans of the first centuries AD, and all those who came within the range of their cultural influence, both human work and the natural world were consecrated to the classical divinities of Olympus and to the spirits of nature. Every place and species, every craft and profession had been assigned its appropriate god or goddess. This order of things was swept away by the Christian revolution on the ground that it was idolatrous, but within a few centuries much of it had been restored under a new guise. The Roman empire may have been deliberately created by divine providence to ease the transmission of the Christian gospel. But the gospel had to adapt itself to the forms of the religious culture which it displaced as the price of that transmission. The reconsecration of nature, the crafts and professions under the new order of Christendom is the subject of this chapter, so far as it applies to the theory and practice of mining and metallurgy. To this end we consider three different ways by which the Church made the metals and metal working holy. The first of these was the attribution of patron saints to the mining and metallurgical professions. The second was the development of a theory concerning the efficacy of the precious stones and metals as aids to the contemplation of God in church worship. The third was alchemy. I have listed the three ways in this order partly for chronological reasons and partly because this ordering is a measure of their relative importance. But each of these ways of consecrating the work and product of mining and metallurgy was a matter of dispute within Christendom. There was often a party within the Church which believed that such consecration was a profanation of Christian worship, a turning away from the one transcendent God to whom alone devotion should be given.

THE SAINTLY PATRONAGE OF MINING AND METALLURGY

By the end of the middle ages all the crafts and professions had been dedicated to the saints, the patriarchs and angels, who seem to have stood in much the same relation to these activities as the Greek and Roman gods and goddesses before them. The process by which these dedications came about and the dates at which they occurred are almost entirely unknown. Similarly it is only with great difficulty that we can establish exactly what this dedication and patronage meant in practice. The special problem of the saintly patronage of the crafts and professions is part of the larger problem of the cult of the saints in general. For though the veneration of holy men and women is common to many religious traditions, the forms which this veneration took in Christendom were in some ways extraordinary, and these forms were always open to the attack that they constituted the paying of those honours to others which were due to God alone. It is difficult to find adequate grounds for the cult of the saints in scripture. The nearest approximation to it is in the apocryphal book of Maccabees where Jeremiah and Onias are seen in a dream invoking with outstretched hands blessings on the Jewish people. St Paul talks of the mystical body of Christ and of the saints as members of the household of God. But all this is too slender a basis for the theory of invocation and intercession by which those on earth could plead with the martyrs, and with other holy people who had died, to intercede for them with Christ at the throne of God. The basis for this cult as it was received in the middle ages was not scripture but tradition.

Why did this cult arise in the early church? It appears that the practice developed of conducting divine service at the tomb of a martyr on the anniversary of the martyrdom. This practice was furthered by the growing cult of the relics of those who had suffered for the faith. These practices can be securely dated as early as the second century. By the time of St Cyprian the belief that martyrs and other holy people had the power of intercession after death was already established. This belief was no doubt founded upon the immediate experience of the faithful who must have felt the active benevolence of these dead souls in their own lives. But equally there can be no doubt that this belief, and the practice of invocation which arose from it, were very similar to the relations which obtained between patron and client in the Roman political system. In this system the head men of a noble family disposed of great power over their kinsfolk. Similarly other dependants, even whole provinces, attached themselves to the family and made suit to its leaders for whatever they needed from the higher levels of the administration. The cult of the saints repeated this system of government in heaven. It is notable that the cult of particular saints became part of the religious practice of some of these noble families at an early date.

But if the Roman system of political patronage tells us something about the forms which were taken by the cult of the saints, it does not tell us as much as do those pagan practices of dedication which the cult renewed. The attribution of the pagan gods and goddesses to almost every aspect of the natural and human orders is the prototype of the system of saintly patronage in Christendom. In this way the cult of the saints enabled a severely modified and restricted polytheism within the emerging Christian world view. That the new polytheism of the saints was similar to that of the pagans is obvious from the fact that the saints served as patrons of many of the same things as the pagan deities had before them. But in many cases the ascription of a saint to a craft or profession appears to have been quite arbitrary and adventitious; there was hardly any connection between the work and the saint's life. In these cases, no doubt, the ascription was justified by stories of saintly visitations to those engaged in the craft. This lack of an obvious connection between the saint and the craft shows how strong the demand must have been to appoint such patrons in the first place.

The saint most widely associated with mining is St Barbara who was revered in this connection in eastern Europe from the first millennium and in central and western Europe from the middle ages onwards. The spread and intensity of this cult are astonishing, given the very tenuous relations between St Barbara and the activity which she patronised. She is said to have lived in the third century and to have died a martyr's death at the very beginning of the fourth. She was the daughter of a rich man called Dioscoros, according to the account of her life given in the medieval book of saints called the Golden Legend. Because she was very beautiful,

Dioscoros shut her up in a high tower, and here, it has been inferred, her solitude turned her to the true faith. When her father asked her which of her princely suitors she would prefer to marry, a generous request on his part, she angrily replied that she had no wish to marry at all.

Soon afterwards her father went on a long journey abroad. Barbara came down from the tower and ordered some workmen who were building a bath house for her father to put in three windows, not two. After some hesitation they agreed and when she had blessed the water, it miraculously cured all the sick who believed in her. Shortly afterwards Barbara defaced the statues of the Olympians whom her father worshipped. At this point her father returned, and when he asked her why she had ordered three windows, not two, she replied that they represented the trinity. In a fury he drew his sword to kill her, but she was whisked away and flew to a nearby mountain. Her father chased her there, caught her and threw her into prison. Then he went to the magistrate and begged that she be tortured. But when the magistrate saw how beautiful she was he gave her a chance to recant. She refused, telling the magistrate that the old gods were nothing but stone statues. Then she prayed that those who made and worshipped them be as stony and useless as they.

At these words the magistrate, too, lost his temper and ordered her to be whipped till she was covered in blood. But her wounds miraculously healed overnight. When the magistrate saw her the next morning he bade her thank the pagan Gods for her miraculous recovery. But she replied that it was Jesus Christ who had healed her. The magistrate then tortured and mutilated her, until finally she was led into the town to be executed. But her father, still in a fury, took her away to a mountain to kill her himself. Then Barbara prayed to Jesus to forgive all those who prayed to him in her name. Her father killed her, but as he was coming down from the mountain he was struck by lighting and nothing was left of him but the ashes.

Such is the life of St Barbara. The story of her life in the Golden Legend is full of details which may be found in the life histories of other early saints, though there are few such histories quite as disconnected and haphazard as hers. Nonetheless she was one of the most widely revered saints in the middle ages. In 1969 she was removed from the calendar when it was revised by Pope Paul VI, on the ground that almost nothing was known of her except her name. This does not mean that she is no longer a saint. It means merely that she is not explicitly nominated for a feast day in the new liturgical year. St Christopher suffered a similar change of status at this time. Since the fourteenth century St Barbara Day has been celebrated every December 4, or on the next following Sunday, in the mining communities of Europe. Hundreds and thousands of miners in Austria, Germany, Italy, and parts of Poland, Czechoslovkia, Hungary and other countries annually celebrate the Barbara Festival, whatever their religious affiliations. So writes Cedric Gregory who has made a special study of the Barbara cult. Churches are dedicated to her all over Europe and the near East, some of them dating from 900AD. Hans Holbein painted a picture of her and the standard representation of her in the illuminated manuscripts of the middle ages shows her in the tower in which her story begins.

That tower and, still more, her instructions for the windows of the bath house help to explain St Barbara's patronage of architecture and building. From this point of view her insistence on the three windows makes very good sense since her belief in the efficacy of representing the trinity by these sources of light is in accord with church theories of architectural symbolism. We are to suppose that because the bath house had three windows those who bathed there could be cured of their ailments. But what of mining? So little has Barbara to do with mining that it is tempting to assume that the author of the Golden Legend and his sources had no idea that Barbara was also a patron of miners. The means by which she is brought into connection with mining in other sources are curious. Perhaps the best attested is that she was thought to have power over lightning , given the peculiar death of her father. So it was common to pray to St Barbara during thunderstorms. Very much later, when gunpowder was introduced into mining, St Barbara's power of protecting people from lightning was extended to include the safe use of explosives. But this was a long time after she had already become the patron of mining.

As she died Barbara prayed to Jesus to forgive the sins of all those who prayed to him in her name, and according to the Golden Legend this wish was granted her. This boon conferred upon the name of St Barbara a special power available to all who at the point of death were unable by their circumstances to gain absolution in any other way. Barbara's name became therefore a protection against sudden unrepentant death, and since this was often the lot of the miner, this may have been the reason for their adoption of St Barbara as the patron of mining. Whether it was or not, there seems to have been some unease over St Barbara's patronage of mining in the later middle ages. While in earlier times St Barbara was supposed to have lived in Antioch, Heliopolis, Nicomedia or Rome, her story was shifted to Athens and the great silver mines of Laureion. The revised story tells how she escaped from her prison and was looked after by the miners of Laureion. But when she climbed out of the shaft she was caught and beheaded by her father. This version connects Barbara with mining much more closely.

In the city square of the Austrian city of Leoben is a seventeenth century fountain. Its centrepiece is a statue of St Barbara adorned in miner's garb, wielding a miner's hammer. On graduation day mining engineers from the local university are required to leap across the fountain, climb the statue and kiss St Barbara, to thank her for her help in their studies. Being drunk they sometimes miss their leap and get a soaking. This leap on graduation day towards the statue of the saint is a symbol of that passing from one world to another, from the lay world to the professional, which is sometimes called a 'rite de passage'. We may note also the special meaning of the kiss, given that St Barbara was a type of all those pious virgins who were married to the Church. Hence her disdain for her mortal suitors even though they were princes. But the graduates are not kissing St Barbara as a woman but as the symbol of their profession, arrayed as she is. The chastity of her life history somehow emphasises this. The kiss signifies that they must be as dedicated to their profession as she was to God.

In Germany, too, there are many stories about saintly assistance to miners. In some the saint is St Barbara, in others St Anne, the mother of the Virgin Mary. These stories explain how the mining city of Annaberg came to be named. There is a variety of legends but they all concern one Daniel Knapp, an old and simple miner much given to praying to the Virgin Mary. This Knapp was in difficult circumstances and St Anne appeared to him in a dream to show him where he should dig. Knapp went to Wittenberg and told the Elector his dream. The Elector then accompanied him to the place which St Anne had shown him, together with a large retinue from his court. There Knapp began to dig and soon discovered a wonderful lode of silver. A city grew up around that spot to work the silver, and this city came to be called Annaberg. To this day the miners of Annaberg are called Knappen. Another version tells how it was an angel who appeared to Daniel Knapp, not St Anne. The angel told Daniel to go next day to a nearby forest. There he would find a fir tree higher than all the others, and in its branches a nest full of golden eggs. Daniel went, found the tree but found nothing in its branches. He climbed down, thought about it, and realised that the branches of the angel's message might be the roots. From an angel's point of view the roots of a tree may well appear to be branches. Digging them, he found great silver veins running in all directions. The coins made from this silver were called angel's pennies.

These stories come from the fifteenth century and charming though they are, they establish no clear connection between St Anne or the angel and mining. Knapp was helped because of his devotion to Mary, and by St Anne because she was Mary's mother. In regard to metalworking, however, the case is different. The patron saint of metalworkers in England is St Dunstan. Unlike St Anne and St Barbara, St Dunstan was famous in his own lifetime, played an important part in worldly affairs and was himself an expert in the arts he patronised. Born at the beginning of the tenth century he became a Benedictine monk and was appointed abbot of Glastonbury. He developed the abbey into a great centre of learning but became embroiled in secular politics and had to flee the country. When Edgar became king, he returned and was appointed archbishop of Canterbury. He was Edgar's chief adviser for sixteen years and with Edgar carried through major reforms of Church and state. He ended his years teaching at the cathedral school of Canterbury and died there. He composed several hymns, illuminated manuscripts and was especially skilled in the metal crafts.

The story of St Dunstan is significant in the sacred history of metal since it points to the very close connection between metalwork and the monasteries from the latter half of the first millennium onwards. St Dunstan's work as the abbot of Glastonbury restored a monastic institution which, like many others, had been severely damaged by the raids of Vikings and Normans. But their raids did not prevent the development of the highest skill in artistic metalwork in the monasteries. Glastonbury was no ordinary monastery. Believed to have been founded by Joseph of Arimathea it occupied a central place in the history of English Christianity, and in mythology, too, as one of the sites of the Arthurian legends. Glastonbury was one of the first in a great network of monasteries stretching all over England, at many of which the metals were mined, smelted and worked. Another is Tintern Abbey, made famous by Wordsworth's poem. In many cases the miners and foundries were at some distance from the monasteries which owned them and had therefore to be worked by lay people since the monks had to be in their monasteries at night.

The ownership of these mines and foundries of Christendom by monks established the sanctity of their operation for a thousand years. The fact that metalworking was often the occupation of monks meant that it was a form of work which they depicted in the books produced in their scriptoria. St Dunstan was not only a metalworker; he was skilled in the illumination of manuscripts. In one such illustration, not by St Dunstan but much later from Cologne, we are shown Bezaleel and Aholiab at work on the furniture of the tabernacle, with the ark of the covenant, the stone tablets and the manna in front of them. Bezeleel is represented as a monk of the period in which the picture was made. Through the doorway of the smithy we see the tents of the Israelites and perhaps Moses sitting in the tent nearest to us. In this the late middle ages reassimilated metalworking into the spiritual order of Christianity by reaffirming its importance in the Old Testament. As with the patronage of the saints, this assimilation of the contemporary world to Biblical times was a way by which work was accommodated to the Christian life.

From the opposite point of view it may be that the spirit of renunciation and the transcendence of the divine in the Jewish and Christian traditions were a spur to the development of their intellectual and social traditions. The overwhelming emphasis on the transcendence of God defied and stimulated the faithful to understand how God is all. Any attempt to understand the reach, as it were, of the divine nature required that they relaxed temporarily their exclusive concentration on its unity and transcendence, and began instead to distinguish and discriminate within it. In the mystical tradition of the Jews this exploration of the divine nature produced a scheme which set out in diagrammatic form the various attributes of God. This scheme was called the Sephirothic tree. In the Christian world the same exploration produced the complicated hierarchies of the angels and archangels, as well as the cult of the saints.

We may distinguish here between two different ways of analysing the divine nature. One concerns itself only with that nature in and of itself by a more or less logical disquisition on first principles, and the other proceeds from a sense of how variously God acts in the world. As we have seen in the first chapter, the later Greek theology made use of both kinds of analysis, distinguishing between the superior and intelligible gods and goddesses such as Zeus and Athene, and the inferior or worldly such as Hephaestus. Even here, however, membership in one or other of these two categories of gods by no means excluded membership in the other. Of these same two categories it is quite clear that the system of saintly patronage belonged to the second. It was a way of distinguishing and articulating some of the ways in which God was manifested in the world. The problem was to make these distinctions without impugning the divine unity and transcendence.

By means of saintly patronage the Church reconsecrated human work, and found a way to distinguish and characterise these various superventions of the divine in human affairs. The patronage of a craft or profession by its own particular saint gave those engaged in that work a sense of the special grace which enabled it. At best this patronage emphasised the special character of the work by a religious sanction which conferred upon the work a holiness and propriety belonging to the work in its own nature. The saintly patronage of work followed on from the classical attribution of Olympian patrons to these same kinds of work, and this continuation of the classical tradition ensured that work was sacred in the west for by far the greater part of its recorded history. But unlike the Olympian patronage in its own time, the saintly patronage was sometimes obnoxious to Christians who rejected the cult of the saints as a whole or severely qualified it. Such a rejection all but destroyed the cult in northwestern Europe in the seventeenth century, at which time work was formally desacralised. The consequences of this departure from tradition are discussed in the fifth chapter of this book. It is enough to suggest here that the loss of saintly patronage from the crafts and professions was an enabling condition of the industrial revolution. The new way of treating human work could hardly have developed if it had remained under the protection of the saints. In such ways do the abstruse theological disputes of earlier centuries determine the greatest issues of the present.

On the other hand the very arbitrariness of saintly patronage was, no doubt, partly to blame for its disappearance. In those many cases where the connection between the character of the work and the character of the patron saint was external, incidental and adventitious, the sanctification of the work did not proceed from a realisation of its spiritual inwardness but was conferred upon it from outside. It was made holy by convention, not in the understanding. To make a particular kind of work holy in the understanding, it has to be shown how that work repeats the divine act of universal creation, as Homer showed it when he described Hephaestus in his smithy, or as Moses showed it in his account of the making of the tabernacle. Or the work must symbolise salvation, as the story of the fiery furnace in the book of Daniel showed this of smelting. But in many cases the life of a saint told little of the craft or profession which that saint patronised, and this was the case with St Barbara and mining. In the case of metalworking and St Dunstan, however, metalworkers had an ideal on which to model themselves, an exemplary figure who had himself lived their life.

PRECIOUS METALS IN CHURCH WORSHIP

> To me, I confess, one thing has always seemed preeminently fitting: that every costlier or costliest thing should serve, first and foremost, for the administration of the holy eucharist. If golden pouring vessels, golden vials, golden little mortars used to serve, by the word of God or the command of the prophet, to collect the blood of goats or calves or the red heifer: how much more must golden vessels, precious stones, and whatever is most valued among all created things, be laid out, with continual reverence and full devotion, for the reception of the blood of Christ.

These words were written about 1145 by Abbot Suger, abbot of the church of St Denis and previously regent of France. They appear in the book which he wrote at the request of the general chapter of St Denis to justify his administration of the abbey. Suger was one of the most influential churchmen of the middle ages, and his book sets out to explain in detail the revolution in church architecture and ornamentation which he brought about during his abbacy. At St Denis Suger began the Gothic order.

For Suger and the chapter of his abbey, the worship of God required the most elaborate and gorgeous equipment, buildings filled with light, the fullest ceremony. The whole of God's creation was to be enlisted in his worship, and this was the proper use for the most valuable materials. Certainly Suger understood that a saintly mind, a pure heart and a faithful intention ought to suffice for this sacred function, to use his own words. But he believed that homage should also be paid by means of outward ornament. He was acutely aware of those among the faithful who denied themselves and others these outward ornaments, who believed that such things distracted the congregation. In his books he sets himself the task of justifying what he had done as though to defend himself against the charge that he had corrupted the purity of Christian worship. We cannot tell whether his revolution in architecture and ornament succeeded because his arguments were convincing, or because his abbey was beautiful, or for some other reason. What we can do is to examine his arguments for the use of the precious metals in church worship as a way of coming to understand their special place in the history of religion, and particularly in Christianity. For there has always been a resistance to the use of such things in the religion whose founder was committed to poverty and renunciation. On the other hand the use of these materials has been a feature of Christian worship from well before the middle ages.

Suger argues his case in this instance on the basis of Jewish ritual. As we saw in the last chapter, the rituals and artefacts of Jewish worship were prescribed by God to Moses on Mount Sinai, where Moses was told in the fullest detail how to make and use the tabernacle and its various shrines and altars. Since, according to Suger, the blood of the sacrificial victims in Jewish ritual was collected in vessels of gold, the blood of Christ in the holy communion should be treated as well or better. This is his justification for the use of golden chalices and monstrances and for the elaboration of the other vessels of worship. He is fully aware in making this claim that he is in danger of damaging the spirit of the ritual by this emphasis on its outward show, and so he insists that these beautiful instruments of the mass be laid out with all due reverence and devotion. He believes that this way of handling the blood and flesh of Christ will conduce to a proper appreciation of the ritual on the part of the worshippers.

Suger refers to the prescriptions of Exodus and some other passages in the Old Testament in his argument. But he must also have been conscious of that passage in the Epistle to the Hebrews which discussed the relation between Jewish sacrifice and the sacrifice of Christ. In that passage the author of the epistle distinguished as sharply as possible between these two sacrifices, arguing that Christ's sacrifice of himself is the substance of which the animal sacrifices of the Jews were but the shadow.

According to the epistle, the Jewish ritual sanctified merely the flesh and the tabernacle made by hands, while Christ's sacrifice sanctified and perfected the conscience and showed the way to that holiest place which was not to be found in any earthly tabernacle or temple but in heaven itself. And this, I suggested in the last chapter, was exactly what was signified by the rending of the veil of the temple at the moment of Christ's death. According to the epistle, Christ's sacrifice completely eclipsed the forms and instruments of Jewish worship and made them redundant.

What then of Suger's claim that elaborate and costly vessels were preeminently fitting for the performance of the holy eucharist? Suger emphasises that the sacrifice of Christ is much more than the animal sacrifice of the Jews. But where the epistle makes this point by stressing the difference between the carnal and the spiritual, Suger makes it by arguing that the physical instruments of Christian worship should be as much more magnificent than those of the Jews as Christ's blood was more valuable than the blood of sacrificial animals. Now it cannot quite be said that Suger's argument contradicts the epistle. In many respects it reasserts the distinctions which the epistle makes between the Jewish and Christian sacrifices. But the means which Suger adopts to demonstrate these distinctions are clearly not those of the author of the epistle. Suger tried to outdo the Mosaic tabernacle in splendour and costliness as a way of demonstrating the superiority of the Christian dispensation over the Jewish. The epistle merely pointed to the superiority of the spirit over all things made by hands.

Immediately after arguing that the use of gold for the vessels of the eucharist is justified, Suger goes on:

> Surely neither we nor our possessions suffice for this service. If, by a new creation, our substance were reformed from that of the holy cherubim and seraphim, it would still offer an insufficient and unworthy service for so great and ineffable a victim; and yet we have so great a propitiation for our sins.

Having argued that the eucharist requires the costliest materials for its performance, Suger now claims that neither we ourselves nor anything else on earth is adequate to the ritual. Christ is so far beyond us that nothing we can do or use is worthy of him, and yet he died for us. That this was truly what Suger felt and no mere conventional piety is clear from the way in which Suger expresses it. For in suggesting that we would still be inadequate even if we were made from the same substance as the highest ranks of the angels, Suger gives us a real sense of Christ's ineffability and transcendence.

These words of Suger are very much closer in spirit to the Epistle to the Hebrews than what he had written before them. But they raise the possibility that there are no vessels more fitting to the performance of the eucharist than any others, given that everything on earth and above is inadequate to this service. Suger tries to exclude this possibility when he says, soon after, that this service requires more of us than anything else we do, and that since God has provided us with everything in the universe, we should make use of everything to worship him. But these arguments would hardly be enough to meet the powerful counter claim that the transcendence of Christ is far more fittingly approached through a deliberate turning away from the world than through the splendour of earthly things. To this claim, Suger would be hard put to find an answer on the basis of his argument so far. In another of his books, *On the Consecration*, Suger even argues that his use of costly materials in worship was clearly sanctioned by the holy martyrs themselves. Just when he needed a large stock of gold and precious stones for a project in their honour, these valuables were put into his hands providentially and free of charge. 'What else was I to do?' he exclaims, since the martyrs were clearly telling him 'Whether you want it or not, we want the best.'

As a former regent of France and the abbot of its richest abbey Suger commanded considerable resources for the implementation of his ideas. So closely was the abbey of St Denis connected with French royalty, and so intimate was Suger himself with the kings of France, that we may suspect the Gothic style, at least as Suger developed it, of being essentially aristocratic. In this way it departed from the Romanesque style which preceded it, where solid walls and simple arches realised the ideals of priest and peasant alike. But for all his power and influence, Suger was not the greatest French churchman of his day. That description fits Bernard of Clairvaux, one of the most ascetic Christians of his time, founder of more than one hundred and fifty monasteries and the mentor of popes. The history of the French Church in this period is polarised between these two colossi, Suger and Bernard, who realised in their time that tension in Christian thought between the sanctification of earthly things and transcendent asceticism. As we read Suger's justification of what he did at St Denis, it is easy to believe that he was writing with Bernard in mind.

There has been a long debate over the last seventy years concerning the exact relation between Bernard and Suger in matters of art and church worship. At first sight it appears that Bernard was opposed to the revolution in architecture, ornament and ritual which Suger brought about during his abbacy. Bernard was a Cistercian, a member of that new order which was founded at the end of the eleventh century to recreate the strictest conditions of monastic life. Under Bernard's leadership this order expanded rapidly during the twelfth century and acquired for its leader and itself a reputation for great sanctity. Bernard himself was famous for the harshness of his fasting and his capacity to do without sleep, to an extent that seemed hardly human. A prolific writer, he has sometimes been credited with having developed the inner life of the soul in a way unknown before either in Europe or elsewhere. The range of emotions which he explores in his spiritual searchings, his hymns and prayers is unprecedented. If Suger was a master of outward form, Bernard was a genius of the inner life.

The Cistercian order was exceptionally austere. It avoided everything like splendour in its churches. Wooden crosses were the rule, there was to be no figural painting or sculpture, no costly vestments, the candlesticks and censers were to be of iron. For the sake of Christ Bernard deemed as dung whatever shone with beauty, enchanted the ear or delighted with fragrance. He hated whatever might distract the attention of the worshipper from inward contemplation, and in a famous passage excoriated the elaborate carvings of monsters and grotesques which were a feature of the great abbey at Cluny, one of the most powerful monastic foundations of the time. When we set this side by side with the ornaments of Suger it is hard to see these two great abbots as anything but bitterly opposed in their theories of religious art. Where Bernard with his wooden crucifixes composed hymns on different parts of Christ's crucified body, Suger contrived his great cross of gold and loaded it with pearls and gems. He justified this by the words of the apostle St Andrew, as he believed, who had described the cross of Golgotha as being adorned with the members of Christ 'even as with pearls.' Where Bernard's rule specified that the communion chalice should be made of materials no more precious than silver or silver gilt, Suger had one made of which the base was gold studded with gems and the bowl was carved from a single, huge sardonyx. This cup has been preserved to our own time.

But Bernard and Suger were friends. The only issue on which we know them to have disagreed was the lack of monastic discipline at St Denis, for which Bernard blamed Suger but for which Suger's predecessor as abbot was almost certainly responsible. In any case Suger seems to have satisfied Bernard in this regard by his reform of the abbey, and from this time onwards the two were in perfect amity. What then are we to make of Bernard's theory of art? Once the issue of monastic discipline is set aside, it is arguable that Bernard never intended his rules for religious art to have general application. They were for the Cistercian order only. If this is so, he may even have approved of Suger's renovations at St Denis. Some scholars have gone so far as to suggest that Suger's revolution is actually the realisation of Bernard's theory of religious art, on the ground that Suger's style rejected the Cluniac order of monsters and grotesques which Bernard condemned. Compared to that order, Cistercian architecture and the abbey of St Denis, as Suger rebuilt it, have in common a spaciousness and calm and absence of crowding. Bernard believed that all the elements of church architecture should be integrated in a single scheme, and in this too he was at one with Suger and the Gothic architecture. Suger was creating a new national French style as was entirely appropriate to an abbey closely connected with the history of the French royalty and which served as the home of the oriflamme, the banner of France. Possibly through Suger's influence, Bernard was himself brought into direct and official relations with the French throne.

There is a further aspect to Suger's theory of art which may well have appealed to Bernard. The abbey of St Denis was named after the man who had converted France to Christianity and whose writings were believed to have been placed in the abbey's library. There they had been translated from Greek into Latin by John Scotus Erigena in the ninth century, and much of what Suger says about art can be traced back to them. This St Denis as he was identified in Suger's time was in fact a composite of three quite different people, since the founder of Christianity in France was not the author of the writings, nor was the author quite whom he seemed to suggest that he was. The author was taken to be the Dionysius who was named in the Acts of the Apostles as Dionysius the Areopagite, the disciple of St Paul. In fact the author was almost certainly a Syrian disciple of the Platonist philosopher Proclus and lived in the fifth century. He had converted to Christianity and transformed Christian teaching by reinterpreting Jewish and Christian scriptures in Platonic terms. This was the man on whose writing Suger drew for his theory of art and whom Suger and his contemporaries mistakenly identified both with the apostle of France and with the disciple of St Paul.

This welter of errors had powerful consequences. Because these writings in the abbey of St Denis were believed to have come from the earliest period of Christian thought, they were accorded a veneration equivalent almost to that accorded to the scriptures themselves. In this way the Platonist teaching of Pseudo Dionysius, as he is now called, entered the mainstream of Christian thought beyond suspicion. Insofar as Suger could draw upon these writings to justify his artistic revolution, he had a warrant for what he did. Art historians in this century have therefore been especially concerned with the connections between Suger's writings on art and the theology of Pseudo Dionysius. It is in this theology, historians believe, that Suger found his arguments for the use of the precious metals and stones in church worship.

There are several passages in John Scotus Erigena's translations of the Pseudo Dionysian writings which historians believe to have influenced Suger in the use of the precious metals and stones. Pseudo Dionysius developed a theory of light:

> Every creature, visible or invisible, is a light brought into being by
> the father of the lights... This stone or that piece of wood is a light
> to me... For I perceive that it is good and beautiful; that it exists
> according to its proper rules of proportion; that it differs in kind
> and species from other kinds and species; that it is defined by its
> number, by virtue of which it is 'one' thing; that it does not trans-
> gress its order; that it seeks its place according to its specific grav-
> ity. As I perceive such and similar things in this stone they be-
> come lights to me, that is to say, they enlighten me. For I begin to
> think whence the stone is invested with such properties...; and soon,
> under the guidance of reason, I am led through all things to that
> cause of all things which endows them with place and order, with
> number, species and kind, with goodness and beauty and essence,
> and with all other grants and gifts.

In such passages, it is supposed, Suger found his justification for the
splendour of his new order. The stained glass, the lustrous metals and
gems provided these 'lights' which Pseudo Dionysius perceived in the world
around him.

It is immediately noticeable, however, that this passage does not dis-
cuss the physical light which illumines and is reflected by the objects which
it describes. Indeed, those objects, a stone or a piece of wood, seem to
have been chosen as examples just because of their lowliness, simplicity
and lack of lustre. They are lights not because they are lit or shine but
because they instantiate certain intelligible properties. Each is one thing
and, in being one thing, realises at its own level the property of unity which
is not a physical but an intelligible property. We do not see that a thing is
one thing with our eyes. We judge it to be so by an act of the intelligence.
And so with the other properties which the passage mentions. The kind or
species of a thing is likewise an intelligible property according to the Platonic school, since each physical object owes its nature as a particular kind
of thing to the idea or form in the mind of God which it copies or in which
it participates. By virtue of its imitating or participating in that idea it is
the kind of thing it is, and reminds us of that idea. In this way the contemplation of things in the world leads us back to the ideas beyond time and
space which make them what they are. Like the picture of someone we
know, they remind us of ideas beyond themselves, and our being reminded
by them of these ideas enables us to recognise them for what they are. As
it stands, therefore, this passage hardly justifies Suger's new order. Those
who believe that it does have perhaps placed too much weight on the author's use of the word 'light' which in this context has nothing to do with
physical light.

Here is another passage from the Pseudo Dionysian writings which is
often adduced to justify Suger's use of the precious metals and stones:

> The material lights, both those which are disposed by nature in the spaces of the heavens and those which are produced on earth by human artifice, are images of the intelligible lights, and above all of the true light itself.

This passage refers to the stars and seems to give support to the view that its author is concerned with visible light. On this view it is the shining of the stars which makes them images of the intelligible lights. But it is unlikely that this is what the author meant. For the Platonist the heavenly bodies are images of the intelligible lights not because they shine but because their positions and movements realise in the best possible way the most perfect mathematical principles and ratios. In a remarkable passage of the Republic Plato actually writes that the true astronomer does not look with his eyes at the heavens at all, but pursues his science by pure reasoning only. That this is what Pseudo Dionysius had in mind is suggested by his reference to the disposition of the stars in the spaces of the heavens, and by his turning from them to works of human artifice as other images of the intelligible lights. These works of human artifice are not particularised as shining or lustrous. They are images of the intelligible lights for the reason I have given at the end of the second chapter. Every work of craft is an image of that idea in the mind of God which the craftworker must contemplate if it is to be realised on earth through the act of manufacture.

There is good reason to suppose that Suger himself understood the Pseudo Dionysian writings in exactly this way. In a poem inscribed on the western portal of St Denis, the doors of which exhibited in relief the passion and resurrection or ascension of Christ, Suger wrote:

> Whoever you are, if you seek to extol the glory of these doors,
> Marvel not at the gold and the expense but at the craft of the work.
> Bright is the noble work; but, being nobly bright, the work
> Should brighten the minds so that they may travel, through the true lights,
> To the true light where Christ is the true door.
> In what manner it be inherent in this world the golden door defines;
> The dull mind rises to truth, through what is material
> And, in seeing this light, is resurrected from its former submersion.

This poem of Suger verbally echoes the passage from Pseudo Dionysius just quoted and makes it certain that Suger was deriving his theory of art from this source. The placing of the poem at the main entry into the church was intended to instruct those who entered on how to understand the works of art which the church contained and organised. Once again the light which the church manifested was not physical light. Those who entered were not to admire the gold or the cost of the doors but the work in them. It was this that was bright, because it manifested the ideas after which the doors were made, the ideas of which the physical works of art were images. Through looking at these physical works and remembering Suger's poem the worshippers were led to a contemplation of intelligible ideas or true lights. This contemplation would lead them to Christ himself, the true light of which Pseudo Dionysius had written.

That this is what Suger meant by his poem on the doors is confirmed by some of his other poems in the church and by numerous passages in his books on his administration and consecration of the abbey. Sometimes Suger seems to catch up with himself as he is carried away by his descriptions of the precious materials he used, stops, and reminds himself and his readers that it is the work which matters and not the value or beauty of the materials. On one occasion he uses a tag from the Roman poet Ovid 'Materiem superabat opus', the work surpassed the material. In this way he was able to defend his revolution in architecture and ornament even against Bernard's rejection of whatever shines with beauty. It is not the shining that matters but the craft, which leads those who see it to contemplation of those ideas from which the craftworkers derived their inspiration.

I have made much of this point about craft not only because Suger does, but because it shows that at this level Pseudo Dionysius gives no warrant for Suger's use of precious materials. The material lights could as well have been made of much humbler materials. In another book Pseudo Dionysius himself actually argues that the divine nature is much better represented by ugly and inharmonious images than by beautiful ones. In this passage he mentions the highest orders of the angels whom he calls the celestial intelligences.

> Nor, I suppose, will any reasonable man deny that discordant figures uplift the mind more than do the harmonious, for in dwelling upon the nobler images, it is probable that we might fall into the error of supposing that the celestial intelligences are some kind of golden beings, or shining men flashing like lightning, fair to behold, or clad in glittering apparel, raying forth harmless fire, or with such other similar forms as are assigned by theology to the celestial intelligences. But lest this thing befall those whose mind has conceived nothing higher than the wonders of visible beauty, the wisdom of the venerable theologians, which has power to lead us to the heights, reverently descends to the level of the inharmonious dissimilitudes, not allowing our irrational nature to remain attached to those unseemly images, but arousing the upward turning part of the soul, and stimulating it through the ugliness of the

images; since it would seem neither right nor true, even to those who cling to earthly things, that such low forms could resemble those supercelestial and divine contemplations. Moreover, it must be borne in mind that no single existing thing is entirely deprived of participation in the beautiful, for, as the true word says, all things are very beautiful.

That Suger understood this to some extent is shown by the last line of one of his poems where he declares that in religious art what is signified is much more important than what signifies it.

So far, then, we have failed to find any adequate justification in Suger's books and poems for the use of precious materials. Instead, in the writings of Bernard and Pseudo Dionysius we have found reasons for excluding these materials from church worship. But when we remember how the precious metals and stones were described in St John's account of the heavenly Jerusalem at the end of Revelation, and when we consider the beautiful reliquaries and church ornaments which were created out of precious metals from the beginning of the middle ages and long before, we are compelled to acknowledge that these stones and metals have a place in Christian worship, though we have so far failed to explain it. Plato too used the precious stones and metals as an element in his account of that real world to which the souls of the dead ascend. We must admit that in the Christian and Platonic traditions as well as in the Jewish, these materials, or at least the idea of them, played a part.

To explain why this is so requires that we engage in the study of what may be called the metaphysics of light at a much deeper level than we have so far attempted. Let us consider for a moment the Platonic theory of ideas again. According to this theory there are beyond this world of space and time a range of principles which are the originals of all the kinds of artefact and species of creature on earth. This range of principles includes also certain fundamental notions such as unity, difference, the terms of the mathematical sciences, and also certain moral and aesthetic principles such as justice, beauty and goodness. Unlike the things in this world, these principles are unchanging and inviolable, to be perceived not by the senses but by the intellect. They condition all our thinking whether we are consciously aware of them or not. Since they are perfect and unchanging and are perceived by the intellect alone, they are said by Plato to excel in clarity and brightness anything and everything in the world of time and space. For this reason they are symbolised by the precious stones and metals in the passage from Plato to which I have just referred. And it is precisely this property of clarity or brightness, with which Suger and Pseudo Dionysius are concerned in their accounts of the material lights. In respect of this property their theories are entirely Platonic.

As Pseudo Dionysius contemplates the stone, he explains what he means by saying that material things, even the most humble, are lights to him. When we looked at this passage before, it was to see how these material lights were not lights by virtue of their physical illumination but because of their intelligible properties. Pseudo Dionysius notes that the stone differs in kind and species from other kinds and species. When he thinks of the stone as a stone, he thinks of all those kinds and species as well, so that the thinking of the stone as a stone brings with it some awareness of the entire range of kinds and species of which the idea of the stone is but one member among others. This is what is required of anyone who is to think of a stone as a stone whether or not the thinker is aware of thinking of all these other kinds and species at the same time. The kinds and species of all created things comprise a single wonderful scheme of which the taxonomies of botany and zoology, of astronomy and mineralogy are parts, as are all the sciences of manufacture. All this, in some sense and to some degree, is before the mind of anyone who thinks of a stone as a stone though most of us are hardly aware of it as we go about our daily lives. For Plato and Pseudo Dionysius wisdom consists in turning round upon ourselves and coming to realise the vast, hierarchical schema of ideas which is implicit in even the least of our thoughts.

And so with the other properties which Pseudo Dionysius ascribes to the stone. To think of the stone as seeking its place according to its specific gravity brings before the mind the system of the elements fire, air, water and earth. For to think of anything as belonging to any one of these, as the stone is of earth, is necessarily to think of all four elements. No one of the elements can be understood except in its relations to the other three. Pseudo Dionysius would have known the passage in Plato's Timaeus where the four elements are shown to be bound to each other by a single ratio. On this account God began his creation with the elements of fire and earth, since the world he was to make had to be visible and tangible. In order therefore to bind fire and earth as closely together as possible, he created between them two means, air and water, so that fire was to air as air to water and water to earth.

With the elements, with number, with the kinds and species, everywhere we see this same rule of interconnectedness. To think of any one term within each of these ranges requires that to some degree we bring before the mind all the other terms within that range. Each term leads us to the contemplation of the entire scheme to which it belongs. It is as though within each term the other terms of the range to which it belongs are all implicit. They are bound together so closely by these necessary relations that Plato and Pseudo Dionysius supposed that finally they are integrated by a principle of unity still greater and more real than themselves. This principle of unity is on another level again from the ideas and essences which it interconnects. It is to this principle that each range of ideas ultimately owes its unity, and at the higher levels these different ranges are themselves necessarily coordinated in yet greater and more inclusive unities, which are finally integrated in one greatest unity which includes them all. In this way, from the contemplation of a stone or piece of wood, the mind of the contemplative is led upwards to this ultimate principle which, by imparting its own unity to all beneath it, accomplishes all the systematic organisation of the world and of our thinking about it. This superessential unity is the ultimate principle of the intelligible world beyond time and space and of the intellect which conceives of the intelligible and of all its relations in this light. This light of the intellect is within each of us; for Pseudo Dionysius and for other Platonising Christians like Suger, it is the true light of Christ himself.

From the poem inscribed on the western portal of St Denis we may be sure that Suger was following Pseudo Dionysius when he identified Christ with the superessential One of Platonism. He is likely to have been influenced also by St Augustine who propounded a similar theory, and by the first verses of the Gospel of St John where Christ was identified with the light of the world by whom all things were made. From at least the end of the thirteenth century it was customary to read these verses of St John's Gospel at the end of every mass. From Pseudo Dionysius Suger would have learnt that not only Christ but the highest orders of the intelligible orders below Christ were divine intelligences. These are the nine orders of the angels, the first receptacles, as it were, of the outpouring of the intelligible light from Christ himself. Pseudo Dionysius carefully discriminated between the different powers of these nine angelic orders, dividing them into three groups of three such that the powers of the first, middle and last of each group marvellously reflected the powers of those angelic orders in the same places in the other two groups. In this way the nine angelic orders were bound together in a mathematical unity.

These angelic orders are the first diversification of the supreme intelligible unity which is Christ himself. From them proceed in turn the different ranks and hierarchies of the intelligible essences which generate the world of time and space in their image. The nine angelic orders stand around the throne of heaven singing in praise of the supreme, whose power proceeds as from a centre through the angelic hierarchies and the intelligible essences into the world around us. Every church is likewise the centre of the community around it, and the centre of every church is the altar from which proceed the four directions along the lines laid out by the cruciform structure of the building. At the very centre the altar represents the supreme principle from whom all things proceed. To walk from the door of the church towards the altar is to go to the centre of the universe, where the highest orders of the intelligible world stand round the throne of God singing in praise. The physical geography of church and altar represents the relations between the supreme principle and everything which proceeds from that principle, even though that principle is beyond the physical world. In the same way the decorations and ornamentation of church and sanctuary can or should represent the invisible nature of that principle in a way which is at once appropriate to it and which renders it apprehensible to us as creatures in the world.

This, we may suppose, is what Suger sought to achieve by his use of precious metals and stones, and by his greatly expanded use of stained glass. These materials all share a capacity to conduct or reflect light and to diversify it. At the physical level they permit the radiation, transmission, refraction and reflection of light as no other materials can. Plato, St John and Pseudo Dionysius used light as the symbol of intelligibility because its pervasiveness, penetration, mingling and reflection best represent the universality and interconnectedness of the intelligible orders. The different gems which served as the foundations of the walls in the heavenly Jerusalem, the gold pavement transparent as glass, the illumination of the whole by the lamb of God, provide an analogy for the relations between the supreme principle and the next highest orders of the divine. The sanctuary of every church is a representation of the heavenly Jerusalem. A similar representation of the divine hierarchies may have been the purpose of the gold lampstand in the holy place of the tabernacle. This lampstand was made from a single ingot which suggests the unity and integrity of the supreme principle. Its seven lamps symbolised the primordial diversification of that unity into the seven planets or the seven days of creation. Once again we find the most precious metal, the divine unity and integrity, its procession into the first plurality and the use of physical light.

The use of light as a means of leading the worshipper to a contemplation of the divine intellect is very ancient. Even in the Greek tradition it is possible to date it back much earlier than Plato, for whom light was the central metaphor of the intelligible world beyond appearances. Homer's description of the great palace of king Alcinous with which the first chapter begins, its gold and silver doors, its golden and silver dogs, its gold youths holding torches, is a symbol of the intellect. From this point of view that palace, and the country which king Alcinous governed with his wife, represent the dimension of the human being which coincides with the divine nature. In the country of king Alcinous the gods appeared without disguise as frequent and honoured guests. Abbot Suger described the peculiar elation he felt as he contemplated the altar of St Denis and his great cross. The loveliness of the many coloured pearls and stones induced in him a meditation on the diversity of the sacred virtues and powers, and transferred him to a strange world which was neither in heaven nor on earth.

CHRISTIAN ALCHEMY

The third means by which the arts of working metal were sanctified in the middle ages was alchemy. In 1142 Robert of Chester, an Englishman, completed his translation of an Arabic text on alchemy. This book inaugurated the study of alchemy in western Europe, a study which persisted for several centuries and may even be said to have continued to our own time. The vast literature on the subject, much of it contradictory and even self contradictory, makes it almost impossible to provide a clear account of the beliefs shared by its practitioners. Furthermore many of the most famous authors on the subject indicated that they were carefully concealing vital principles in their writing, as a way of ensuring that their science was not infiltrated by people who were unworthy of it. These writers had good reason to take precautions. Their science, if that is what it was, provided many opportunities for fraud. Chaucer in his *Canterbury Tales* described in detail how the gullible could be tricked into paying large sums for a powder which seemed to convert base metals into silver.

A further obstacle to the understanding of alchemy in our time is our firm conviction that metals cannot be changed into one another by chemical manipulation. But many alchemists claimed that this was not the object of their work. The object of alchemy was to effect a spiritual, not material, transmutation. On this account the transmutation of metals was at most a side effect of spiritual transmutation. Some have argued that the chemical language of alchemy was nothing but a screen to mask a theory of spiritual development, in which the gold which the alchemist set out to make was no more than a metaphor for a spiritual state. It is notable, for example, that one of the greatest churchmen of the twelfth century, Albert the Great, was firmly committed to the theory of alchemy but took no interest in the physical practice of it. On this view the philosopher's stone, that marvellous substance which was credited with the power of transforming base into precious metals, was Jesus Christ himself, through whose operation the leaden and debased soul could be restored to its original gold.

Whether metaphor or not, it was generally believed that the metals were transmutable into each other. One reason for this may have been that the metallurgists of the time found it difficult to distinguish between the appearance and the substance of the metals. Alchemy seems to have derived in part from those arts which could create the appearance of a metal, such as gilding, and from the arts of dyeing and staining glass. From the Arabic texts the medieval alchemists acquired the theory that all metals were the result of an interaction between sulphur and quicksilver; the differences between the metals were the effects of the different proportions and qualities of these two prime substances in their constitution. On this basis they supposed that an alteration in the balance of these properties or in the quality of the sulphur or quicksilver would fundamentally alter the constitution of a metal and transform it into another. Some Arab scientists, such as Avicenna, believed that no transmutation was possible, even though they propounded the theory that all metals were the product of these two prime substances. The two substances were themselves obscure. They were not the sulphur and quicksilver to be found in nature but substances much purer and more powerful than these, especially in gold and silver.

In the case of gold the composition was generally agreed to be the finest and most perfectly proportioned of all. Gold was, as it were, the final state of metal, the point beyond which no improvement could be made. Nature grew gold as she grew and matured the other creatures, and the metals other than gold were often regarded as stages of development on the way to the production of gold. In many cultures it was believed that any metal left long enough in the ground would eventually become gold. The work of the alchemist was therefore a kind of forcing, by which the slow processes of nature were speeded up in the laboratory. Since the metals had a life of their own and were capable of maturation, their natural condition in the earth was compared to that of the embryo in the womb. Similarly the furnace of the alchemist was a kind of womb in which the baser metals would come more quickly to their final state. Some believed that the baser metals were originally gold but had deteriorated like the rest of nature as a result of Adam's fall. The alchemical work was therefore a restoration of these metals to their pristine condition as the good gold of Havilah in the garden of Eden.

Adam's fall, it was believed, occasioned a general deterioration in the state of nature. This is by no means explicit in the work of alchemists generally, but a consideration of it will help us bring to life the peculiar mentality of the alchemist. When Adam fell and original sin was brought into the world, the natural creation was largely corrupted. Adam and Eve were compelled to leave the garden for a wilderness from which they would win a living only by sweat and toil. At this time, it was believed by some, the earthly globe lost its perfect sphericity and the mountains and valleys were formed. The metals were debased and all earthly creatures were precipitated into that state of war in which the lion could no longer lie down with the lamb. All of this was the result of Adam's spiritual corruption, since on this view what happens in the human spirit determines the constitution of the material world. That world is, as it were, an expression or extension of the spiritual state of the human.

Adam's spiritual state after the fall, the condition of every human soul in the state of original sin, was compared to the base metals. If gold was the analogue for the human soul in the state of innocence, the base metals represented our fallen nature. But that nature was not irredeemable, it could be reformed. In order to achieve this, it was necessary first that the corrupted soul be dissolved, that it undergo a kind of death to purify it. The hardened selfishness of the soul which had separated itself from God had to be destroyed, and since all human beings in the state of the original sin identified themselves with their separated and godless natures, the process by which they were saved from this seemed nothing less than death to them. In the case of a base metal, the process by which its restoration to gold was begun consisted in its being broken down into a kind of primal matter in which its determining qualities as a particular metal were completely lost. This process was often represented as a putrefaction, accompanied by the most evil stench and concluding with the production of a black, fluid mass. Some authors suppose that this first stage in the alchemical work was really a descent into Hell or Hades, like the descent of Persephone in the Eleusinian mysteries or the passion of Christ. It was the death which was followed by the initiatic rebirth, simulated in the rituals of many cultures. The alchemy practised in the laboratory made metallurgy into a ritual.

The first stage of the alchemical work was signified by the colour black; the next by the colour white. When the black, putrid mass of the dissolved base metal had been properly prepared, it would transform itself after a certain time into a white shining substance of extraordinary purity and brilliance. This white substance is called by a great number of names in the literature, of which the most important is 'quintessence'. Like the dissolved base metal at the black stage, this quintessence could not be said to be any of the metals or even the elements, but was instead the fundamental substratum from which the elements and metals were formed. It stood, therefore, beyond determinable formations, and was the medium or prime matter out of which all things were made. It was the fifth element, ether, of which the other four elements were derivations. It was pure potentiality and contained, as it were, the whole of the creation which it would bring into existence when acted upon by the informing principle. In this state of pure potentiality and passivity, it was sometimes called the lower waters from which God divided the upper waters at the beginning of the creation story in the first chapter of Genesis. It was also called the pure quicksilver, not the quicksilver or mercury found in nature which was often used as the dissolvent of the base metal in the first stage, but something much finer than this, the very principle of quicksilver. In this respect it was the complement and opposite of sulphur, the other of the two prime substances from which all the metals were made. Every base metal contained some sulphur, but by the process of purification the binding, hardening, formative power of the sulphur in a base metal was removed, leaving the quicksilver in its purest form.

To return to the soul. Just as the base metal was dissolved, so the soul died to itself, and after suffering the horrors and torments of this death it was ready to be made anew. At this point, and before it was reformed, it was in a state of pure passivity, receptive to any impression or character which the spirit might give to it. It had now escaped from the limitations of creaturely consciousness and had the potential to achieve a quite different state of being. It was, as it were, virgin, and was compared to the condition of the mother of Jesus at the annunciation. According to the theologians of the middle ages, Mary was at this stage absolutely empty, unsullied by any selfish tendencies, a pure vessel for the Lord. This they took to be the meaning of the virgin birth. For whenever a soul is found in this state of emptiness, God must enter into it. In this way the first beatitude was understood:

> Blessed are the poor in spirit, for theirs is the kingdom
> of heaven.

Poverty of spirit was this emptiness of the soul after escaping the limitations of our individual nature. So poor indeed are the poor in spirit that they do not even know their own poverty. They can neither be said to be, nor do they want to be. They are completely dead to any tendency from their former lives.

The third and final stage of the alchemical work was signified by the colour red. The pure white quicksilver was now recombined with the pure sulphur, which imparted to it the solidity and stability of a metal, and also a certain ruddiness which was more marked in the case of gold than of silver, but was to be found in both. This reunion of the sulphur and quicksilver, which had been combined in the base metal before its dissolution, was called the alchemical marriage. It was regarded as the uniting of spirit and soul, or form and matter. In this marriage the sulphur was regarded as male and the quicksilver, the unformed matter, as female. The gold and silver produced from their union were therefore the result of the perfect reintegration of the sexes. Similarly the sulphur represented the sun and the quicksilver the moon, and their union the moon's illumination by the sun's rays. In these ways the final stage of the alchemical work brought together in perfect harmony and balance the two great energies of the universe, the two predominant forces into which the principal unity had been divided in the beginning, the upper and lower waters. All created things were combinations of these two forces, and alchemy was the art by which they could be disjoined and recombined in exactly the way they had been before the fall.

The pure white quicksilver at the second stage of the alchemical work represented the condition of the soul from which all selfish tendencies had been removed. Its equivalent in the Christian story was the condition of the virgin Mary at the time of the annunciation. Similarly, the recombining of this pure quicksilver with the pure sulphur was symbolised by the descent of the holy spirit upon Mary, as the angel of the annunciation foretold. As a result of the descent Mary conceived. In Christian religious art the descent of the holy spirit, here as elsewhere, was represented by a descending dove, and this is what we find in a picture of the alchemical marriage from a work by an alchemist of the thirteenth century. From this point of view Jesus himself was the product of the alchemical marriage, and this gives a special meaning to the first of the three gifts of the Magi. The presentation of gold to Jesus at the time of his birth may signify the divinity or kingliness of Jesus. But from the alchemical point of view it realised the principle of 'like to like'. Jesus himself was the gold generated by the alchemical work, and consequently gold was the first gift he was given.

So much for the alchemical theory of Christian alchemists in western Europe during the middle ages. We can now give some sense to their claim that the philosopher's stone which transmuted base into precious metals was Jesus Christ. Jesus was the second Adam who showed how the sinful nature which we have inherited from the first Adam can be overcome. For the alchemist this redemption was achieved through the alchemical work in the course of a life on earth, and not after death. Not only was the soul of the alchemist saved by the successful completion of the work, his body was too. The entire person was made anew. The reward which alchemy promised its practitioners was therefore paradise on earth, a terrestrial paradise, a return to the garden of Eden and to the state of things before the fall. This restoration of fallen human nature would occasion a similar restoration in the rest of fallen nature, and with it the transmutation of the base metals into the precious metals from which they had degenerated.

At this point we may detect in the theories of the alchemists a distinctly prechristian strain. The regaining of the terrestrial paradise was not regarded as the final goal of the Christian life. That goal was the gaining of the celestial paradise after a successful passage of the last judgement. In this way alchemical theory looks back to our Adamic origins rather than forward to the life beyond death. When we recall that the story of Adam was even more central to Judaism and Islam than it was to Christianity, the possibility arises that western alchemy, for all its Christian symbolism, had more in common with these traditions than with Christian doctrine.

This becomes even more plausible when we take into account the characteristic doctrines of the Jewish Cabbala, that body of Jewish mystical writing in which the perfect human being, Adam Kadmon, is described. Then there is the story of the fiery furnace in the book of Daniel, where Daniel's three companions pass unscathed through the flames. Here, if anywhere, was the scriptural authority for assimilating the process of human perfectibility to metallurgy, and once again it was a Jewish scripture, rather than a Christian. So we find in alchemy what we have found in the patronage of the saints and in the writings of Abbot Suger: an attempt to reintegrate the divine and the earthly at a level which the Christian emphasis on the divine transcendence did not easily admit.

Did the alchemists make gold? To this question the first answer must be that the belief that they could make gold was part of a much larger structure of beliefs concerning the power of the human spirit to affect the material world. In the case of alchemy this structure of beliefs involved the reading of both Old and New Testaments in a special way, and an acceptance that there were certain profound analogies between states of the soul and the metals. The conviction that a fully realised human being could directly affect the material realm by the power of faith was the basis of belief in Christ's miracles, many of which involved transubstantiation. With Christ's miracles as with alchemy, we are confronted with a world in which the material realm is far more permeable to the spirit than we now suppose it to be. The alchemist aimed to recover the state of Adam and Eve before the fall. Just as the fall brought about a degeneration in the rest of nature, so the spiritual return to paradise would be accompanied by a natural regeneration. At the theoretical level this was quite straightforward, and insofar as it encouraged a belief in the omnipotence of the spirit, it had value regardless of what might be or was actually achieved. We conclude with an abbreviated autobiography of a most famous western alchemist:

> I, Nicholas Flammel, Scrivener, living in Paris, in the year of our Lord, 1399, in the Notary-street, near St James, of the Boucherie, though I learned not much Latin, because of the poverty of my parents, who, not withstanding were, even by those who envy me most, accounted honest and good people; yet, by the blessing of God, I have not wanted an understanding of the books of the philosophers, but learned them, and attained to a certain kind of knowledge, even of their hidden secrets. For which cause's sake, there shall not any moment of my life pass wherein, remembering this so vast good, I will not render thanks to this my good and gracious God.
>
> After the death of my parents, I, Nicholas Flammel, got my living by the art of writing, engrossing, and the like; and in the course of

time, there fell by chance into my hands a gilded book, very old and large, which cost me only two florins. It was not made of paper or parchment, as other books are, but of admirable rinds, as it seemed to me, of young trees; the cover of it was brass, well bound, and graven all over with a strange kind of letters, which I took to be Greek characters, or some such like. This I know, that I could not read them; but as to the matter which was written within, it was engraved, as I suppose, with an iron pencil, or graver, upon the said bark leaves; done admirably well, and in fair neat Latin letters, and curiously coloured. It contained thrice seven leaves, for so they were numbered on the top of each folio, and every seventh leaf was without writing; but in place thereof were several images and figures painted.

At length, after twentyone years of study and fruitless toil, their meaning was explained to me by a Christianised Jew, whom I met with in my travels to discover the meaning of the book. On my return home, I set to work and succeeded in the discovery. He that would see the manner of my arrival home, and joy of my wife Pernelle, let him look upon us two in the city of Paris, upon the door of the chapel of James, in the Boucherie, close by one side of my house, where we are both painted, kneeling, and giving thanks to God: for through the grace of God it was, that I attained the perfect knowledge of all that I desired. I had now the prima materia, the first principles, yet not their preparation, which is a thing most difficult above all things in the world; but in the end I had that also, after a long aberration and wandering in the labyrinth of errors, for the space of three years. During which time, I did nothing but study and search and labour, so as you see me depicted without this arch, where I have shown my process, praying also continually unto God, and reading attentively in my book, pondering the words of the philosophers, and then trying and proving the various operations which I thought they might mean by their words.

At length, I found that which I desired; which I also soon knew, by the scent and odour thereof. Having this, I easily accomplished the magistery. For knowing the preparations of the prime agents, and then literally following the directions in my book, I could not then miss the work if I would. Having attained this, I came now to Projection; and the first time I made projection, was upon mercury; a pound and a half whereof, or thereabouts, I turned into pure silver, better than that of the mine; as I proved by assaying it myself, and also causing others to assay it for me, several times. This was done in the year A.D. 1382, January 17th, about noon, in my own house, Pernelle alone being present with me. Again following the same directions in my book, word by word, I made projection of the Red Stone, on a like quantity of mercury, Pernelle only being present, and in the same house; which was done in the same year, April 25th, at five in the afternoon. This mercury I truly transmuted into almost as much gold, much better indeed than common gold, more soft also, and more pliable. I speak in all truthfully. I have made it three times with the help of Pernelle, who understands it as well as myself; and, without doubt, if she would have done it alone, she would have brought the work to the same, or full as great perfection as I have done. I had truly enough, when I had once done it; but I found exceeding great pleasure and delight in seeing and contemplating the admirable works of nature within the vessels. And to show you that I had then done it three times, I caused to be depicted under the same arch, three furnaces, like to those which serve for the operations of the work.

I was much concerned for a long time, lest Pernelle, by reason of extreme joy, should not hide her felicity, which I measured by my own; and lest she should let fall some words amongst her relations, concerning the great treasure which we possessed. But the goodness of the great God had not only given and filled me with this blessing, in giving me sober chaste wife; but she was also a wise prudent woman, not only capable of reason, but also to do what was reasonable; and made it her business, as I did, to think of God, and to give ourselves to the work of charity and mercy. Before the time wherein I wrote this discourse, which was at the latter end of the year 1413, after the death of my beloved companion; she and I had already founded and endowed with revenues fourteen hospitals, three chapels, and seven churches, in the city of Paris; all which we had new built from the ground, and were able to enrich with gifts and revenues. We have also done at Boulogne about the same as at Paris, besides our private charities, which it would be unbecoming to particularise.

Building, therefore, these hospitals, churches etc., in the aforesaid cities, I caused to be depicted under the said fourth arch, the most true and essential marks and signs of this art, yet under veils and types and hieroglyphical characters; demonstrating to the wise and men of understanding, the direct and perfect way of operation and literary work of the philosopher's stone; which being perfected by any one, takes away from him the root of all sin and evil; changing his evil into good, and making him liberal, courteous, religious, fearing God, however wicked he was before, provided only he carries through the work to its legitimate end. For then thenceforward he is continually ravished with the goodness of God, and with his grace and mercy, which he has obtained from the fountain of eternal goodness; with the profundity of his divine and adorable power, and with the contemplation of his admirable work.

Chapter Five

THE SYMBOL OF THE MINE

This done, when you have a goodly provision of foodstuffs, a goodly number of miners experienced in the work you wish to do, and all the tools necessary for breaking and excavating rocks and carrying away earth, and when you have had the mountains and all your cabins and the mine baptised by a priest in the name of God and a fortunate outcome (dedicating it as is customary to the Holy Trinity or to Our Lady or in the name of some other saint whom you hold in reverence, invoking his protection), then carefully make a beginning of the mining, with the determination to continue and not to abandon the enterprise as long as the possibility of discovery justifies the expense, or until you have passed beyond the confines of the signs shown to you above.

Biringuccio

SO FAR WE HAVE seen more of the furnace and the forge than of the mine. The reason for this is that the ancient and medieval texts have much to say about the smelting and working of metals but little about the mining of them. In the Homeric poems and in the Bible the means by which the metals are taken from the earth hardly figure. Hephaestus proceeds with his work of universal creation by making use of metals already to hand; the vessels and other metal objects of the tabernacle are made from the treasures which the children of Israel took with them from Egypt. In the middle ages, again, the emphasis was on the place of the precious metals in church worship and on the transmutation of one metal into another. Nonetheless the act of mining has its own significance from the spiritual point of view, and this is the subject of this chapter.

Digging into the earth in order to discover and free the metals is a symbol of self discovery. Turning inwards we discover through introspection the kingdom of heaven within us. The treasure in the ground corresponds to the treasure hidden within our own hearts. The darkness of the mine and the difficulty of the miner's work correspond to the otherworldliness and hardships of the spiritual quest. For these reasons the cave in the mountain, like the chamber of the pyramid, is a sacred place in which rites of initiation are performed. In Rome the earliest Christians celebrated their mysteries in the catacombs. Still today there are chapels underground in many mines around the world. Like the spiritual quest the journey into the world beneath the ground is a very dangerous one, and those who are to undertake it require a special preparation. On graduation the mining engineers of Leoben in Austria kiss the statue of St Barbara after leaping across the basin of a fountain. And at Leoben and elsewhere there is another ritual for the initiation of miners and mining engineers, the ledersprung or leap across the leather apron. This ritual too prepares miners for the dangers of their calling.

The leather apron is one of the most ancient and most typical parts of the miner's equipment. It is worn on the back of the body to cover the buttocks, and its German name is arschleder. Black, hairless and cut from cowhide, it protected the miners against damp and enabled them to slide down steep places within the mine. When it is time for apprentices or students of mining engineering to join the society of miners, they are required to undergo a simple ritual of admission. After answering some questions, they are made to stand on a chair, while before them and at a certain height a leather apron is held by experienced miners. Those who hold the apron are usually the most senior administrator present and the oldest miner or mining engineer. The neophyte then leaps from the chair across the apron.

This ritual leap, like the leap across the fountain, symbolises the passing from one world to another across a sacred threshold. It is a moment fraught with power, however hilarious the occasion. The jump requires courage. With both feet at the same time, the miner jumps into the new profession which does not tolerate any stumbling, but demands quick determination and energy. The leaping must be firm if life is to be won. The leapers as they leap are between two worlds. For that instant they are beyond all worlds. Their state is equivalent to the unconditioned spirit which precedes and underlies all the forms and limitations of manifested existence. Their being in the air represents this momentary liberation from the limits of the human, as does levitation. When they alight on the other side they are not the people they were. The old self has been annihilated in the passage of the leap and when they stand on the other side of the apron they are different as their new status and titles indicate. In the case of the miner or mining engineer this difference is very clear. They are committed now to a world quite unlike the world in which most people live, without sunshine, in constant fear of life and limb. To a much greater degree than in most crafts and professions they are now dependent on the skills and science they have learnt. From moment to moment their physical survival will turn on the quickness of their minds and the sureness of their knowledge.

DRAGONS

Perhaps in part because of the special dangers of the mining life, the ceremony of the ledersprung is still performed in central Europe, particularly in Austria, Slovenia, Croatia and Bohemia. Where such rituals have long been lost to other crafts and professions, the miners have preserved theirs. Recently the ritual of the ledersprung has been taken up by the miners of the Ruhr in Germany. This ordeal and what it signifies has much in common with a very widespread motif in folklore and mythology concerning the dangers through which a hero must pass in order to acquire a golden treasure. The earliest European example of this motif is the labour of Hercules to win the golden apples of the Hesperides. These apples were guarded by a dragon with a hundred heads, each speaking with a different voice. This dragon was the child of Typhon, the serpent of the earth goddess which Apollo slew with his bow when he assumed control of the Delphic oracle. If the golden apples correspond to the kingdom of heaven, as gold stands for the realised soul in alchemy, then the dragon which guards the apples stands for the fallen world which must be overcome before the treasure can be won. Its hundred heads and voices represent the multitude of creatures in the world, all of which must be transcended by the aspirant to spiritual fulfilment.

In the story of Hercules the gold grows on a tree in the western isles where the sun sets. It is not found underground. But in the Anglo Saxon story of Beowulf, there are two separate adventures in which the hero acquires an underground hoard by slaying a monster or dragon. In the first adventure Beowulf dives into a pool and after swimming down for a whole day, is dragged by the monster into a dry cave. Here the man destroys the monster, views the hoard and then swims back to his companions. In the second adventure, in which Beowulf himself is killed, he fights a dragon which guards an immense store of ancient treasure in a barrow or burial mound. This treasure had been placed there in another age by the last survivor of a great race. When he died, a burning dragon of the night found the hoard and set himself to guard it. For three hundred years he stayed there, until a foolish slave stole one of the cups and gave it to his master. After waiting for nightfall the dragon took his revenge by destroying people and buildings with his blazing breath. Beowulf's own hall was destroyed, and the old king now summoned his courage for his last trial of strength.

Taking with him some of his earls Beowulf is led by the man who stole the cup to the entrance of the barrow. Just before they reach it, Beowulf tells his companions to wait while he goes on alone to confront the monster. The dragon emerges from the barrow breathing fire. They engage but Beowulf's sword breaks, dealing a less than mortal blow. Beowulf is almost overcome by the dragon's fire and his companions flee. But one of them, Wiglaf, goes to his lord's help and together they kill the dragon. Beowulf has been so savaged by the dragon that he cannot survive. He sends Wiglaf into the barrow to bring the treasure to him so that he may gaze on it as he dies. Within the barrow Wiglaf finds a wonderful hoard of jewels, cups, helmets, arm rings, and above them a standard woven in gold which lights the rest of the treasures. Taking what he can in his arms, he returns to Beowulf who thanks God for the hoard for his people's sake. After making Wiglaf his heir, Beowulf dies. Wiglaf decrees that the treasure shall be burnt with Beowulf on his pyre. For this treasure was hedged about with a spell. No one was permitted to touch it unless God himself enabled one he chose to open the hoard.

In this last adventure of Beowulf the treasure in the barrow is once again a symbol of spiritual realisation. In this case it is acquired by Beowulf at the cost of his life and he wishes to look at it as he dies. It is the promise of his salvation beyond death, and in this respect the story of Beowulf's last moments has a rather more Christian resonance than, say, the doctrines of alchemy which we examined in the last chapter. For there too gold was a symbol of spiritual realisation, but a realisation achievable in this life. The poet of Beowulf makes explicit the connection between the discovery of the hoard and God's will, and the burning of the treasure with the dead king emphasises that the treasure is his and his alone. The spell on the treasure and, even more, the lighting of it by the golden standard shows how its significance transcends the earthly domain. It is not mere wealth, but is charged with supernatural power. In his readiness to fight the dragon at the risk of his life and in his wish to leave the treasure to his people, though this does not happen, Beowulf demonstrates the self sacrifice and love of others which are the true gold of the enlightened spirit. In all this we see how the recovery of treasure hidden underground represents the spiritual quest.

What then of the dragon? In the story of Hercules and the golden apples, the dragon with its hundred heads and hundred voices, child of the serpent of the earth goddess, symbolises the plurality of the created world which must be overcome by the spiritual seeker. And this may be the case with Beowulf's dragon also. But there are other possibilities, consistent with this interpretation but going well beyond it. The dragon of the hoard is an ancient creature, like the hoard which he guards. He has been on guard for three hundred years when the cup is stolen. By overcoming the dragon Beowulf wins the treasure which has been kept from the world for all that time, a treasure which symbolises spiritual fulfilment and is the last relic of an heroic race. Here we may recall the significance of the neophyte's leap in the ledersprung, in which he leaves behind the old self in order to be made anew. The guardian dragon is that old self, the hardened egoic personality which keeps us from our own deepest natures. The dragon is Holdfast, the indurated habits and customs of the selfish soul, the stultifying conventions of the unredeemed. In killing the dragon and recovering the hoard Beowulf reopens the path to our divine destiny and releases the treasures of the spirit. On this account the dragon is not only the created world, but our fallen human nature as well. Like the entry into the tomb, the killing of the dragon is the death of the mortal self to recover the immortal within. Every descent into the mine is such a death.

We have identified the dragon which guards the treasure as Holdfast, the old ways or the old self which have to be discarded to release the spirit. This suggests a further identification between the dragon and the treasure. Are they not the same, just as the unregenerate soul, and that same soul redeemed, are the same soul? The dragon and the treasure are found in the same place to which the dragon is drawn by the treasure; they are both remarkable for the richness, brightness and variety of their coloration; they are both connected with fire and night. This intimate relationship between dragon and treasure is brought out by this passage from *The Hobbit* by JRR Tolkien:

> There he lay, a vast red golden dragon, fast asleep; a thrumming came from his jaws and nostrils, and wisps of smoke, but his fires were low in slumber. Beneath him, under all his limbs and his huge coiled tail, and about him on all sides stretching away across the unseen floors, lay countless piles of precious things, gold wrought and unwrought, gems and jewels, and silver red stained in the ruddy light.

> Smaug lay, with wings folded like an immeasurable bat, turned partly on one side, so that the hobbit could see his underparts and his long pale belly crusted with gems and fragments of gold from his long lying on his costly bed. Behind him where the walls were nearest could dimly be seen coats of mail, helms and axes, swords and spears hanging; and there in rows stood great jars and vessels filled with a wealth that could not be guessed.

Smaug lies with his treasure in the bottommost cellar or dungeon hole of the ancient dwarfs right at the mountain's root. He symbolises the enormous energies of unregenerated nature. In alchemical theory the dragon is the latent power of the base metals, which is destroyed and remade in the processes of the alchemical marriage. In the theory of Yoga, this serpent energy lies coiled at the base of the spine from which it must be led up-wards through the centres or chakras along the spine until it reaches the foramen of the skull. The serpent is potentially therefore the power of the redeemed or realized soul, before that stage is reached. In a similar man-ner one of the early church fathers claimed that the fires of hell were noth-ing more than the visions of the blessed seen from the point of view of the damned. In the story of Beowulf the dragon is described as a 'tangle thing'; in the passage quoted from *The Hobbit* Smaug's tail is coiled. This coiling and entanglement symbolise the confused, unrealised energies of the unre-generate soul.

Another story in which a hero slays a dragon who guards treasure appears in the German Nibelungenlied and the Icelandic Volsunga saga. There are several versions of the story but they all concern the encounter between the hero Siegfried or Sigurd and the dragon Fafnir. This dragon guards the treasure of the Nibelungs and Siegfried fights him at the foot of a mountain where the treasure is hidden. Siegfried is warned to dig a ditch before the fight lest he be overwhelmed by the blood of the dragon. Ac-cording to some accounts Siegfried eats the heart of Fafnir after killing him, and this eating of the dragon's heart makes Siegfried the wisest man in the world. Like the treasure, the dragon's heart is a symbol of spiritual wisdom. But in the Nibelungenlied we are told that Siegfried bathed in the dragon's blood and by so doing made himself invulnerable. This is how Kreimhild, Siegfried's wife, describes it:

> When he slew the dragon at the foot of the mountain the gallant knight bathed in its blood, as a result of which no weapon has pierced him in battle ever since. Nevertheless, when he is at the wars in the midst of all the javelins that warriors hurl, I fear I may lose my dear husband. Alas, how often do I not suffer cruelly in my fear for Siegfried! Now I shall reveal this to you in confidence, dearest kinsman, so that you may keep faith with me, and I shall tell you, trusting utterly in you, where my dear husband can be harmed. When the hot blood flowed from the dragon's wound and the good knight was bathing in it, a broad leaf fell from the linden between his shoulder blades. It is there that he can be wounded, and this is why I am so anxious.

Eventually Siegfried is killed by being wounded in this spot. But for our purposes what matters is the bath of dragon's blood. It is a kind of baptism in which the power of the dead or dying dragon is transmitted to the hero. The invulnerability which it confers is a symbol of spiritual power since the spirit is itself inviolable. In this story we see clearly how the killing of the dragon and the gaining of the treasure both represent spiritual regeneration.

DWARFS

These stories of dragons and treasure signify how the winning of self realisation or redemption is an ordeal through which the hero has to pass. In this respect the stories are comparable to the initiatic ordeals undergone by the graduates of Leoben in kissing St Barbara and in the ledersprung. Since the dragon stories culminate with the gaining of a golden treasure and since that treasure is often underground, it is likely that these stories had a special significance for miners. For the most part the treasure guarded by the dragon was not the unworked metal, still less the ore, but ancient masterpieces of the goldsmith's and armourer's art. But there is another feature of this treasure, as it appears in many of the stories, which enables us to compare it with the miner's product more directly. For very often the treasure guarded by the dragon was made or owned by dwarfs, and dwarfs were peculiarly the supernatural denizens of the mine. Smaug and his treasure lay in the cellar of the ancient dwarfs at the root of the mountain, just as in the Teutonic mythology, the treasure of the Nibelungs belonged first to the dwarf Andvari or Alberich. In this mythology the dwarfs also manufactured the marvellous treasures of the gods, Odin's spear, Thor's hammer, Freyr's ship. The same peoples among whom this mythology was current, developed an elaborate lore concerning the role of the dwarfs in mining.

It would be wrong, however, to suppose that these dragons and dwarfs were exclusively northern European. We have seen at least one southern European dragon who guarded treasure in the Greek story of Hercules and the golden apples of the western isles. Herodotus, the first historian of ancient Greece whose work we have, relates how the Persian king Cambyses conquered Egypt and desecrated the temples of Ptah and the Kabeiroi at Memphis. Herodotus believed that the Greeks had acquired almost all their gods and goddesses from Egypt, and that the Greek Hephaestus was the Egyptian Ptah. Herodotus records that the Ptah whose temple Cambyses desecrated was represented there in the figure of a dwarf, and that the Kabeiroi whom he believed to be Ptah's sons were also dwarfs. These Kabeiroi were primordial workers of metal like their father. There is therefore a close connection between dwarfishness and the working of metals in the Egyptian mythology as well as in the Teutonic, and this presents a problem similar to the one we have already encountered in trying to explain the lameness of Hephaestus. In both the Egyptian and the Teutonic mythologies the arts of working metal were attributed to supernatural beings of the greatest antiquity who were also dwarfs. What is it about these arts which led to this conception? And what is the connection between dwarfs and mines?

If dwarfs were involved only in the mining of metals and not in the working of them, we might explain their size by supposing that it helped them pass more easily through the galleries of the mine. Or we might attribute their smallness to their being a class of elves, except that elves are by no means uniformly small though they are rarely larger than humans. The dwarfs lived in secret places and tended to avoid the company of humans. In this, too, their smallness was an advantage since it enabled them to escape human observation when they wished. They usually lived underground and were gifted with exceptional intelligence. But unlike the other elves, they were rarely beautiful, having big heads, pale faces and long beards. They were often hunchbacked or otherwise deformed. In German folklore they usually wore brown or black frocks, leather belts and high pointed hoods, clothes so like those of the traditional miners that it is often difficult to tell whether a picture of a miner is a picture of a dwarf or the other way around. Nonetheless, despite their shortness, age and garments, they were great dancers, like the other elves. They were usually found in groups of which seven and thirteen seem to have been the most common numbers.

It is difficult to determine how popular beliefs in elves and dwarfs coexisted with Christian doctrine in the German mining communities. Many of the best stories about miners and dwarfs come from the sixteenth century, a period of religious enthusiasm in this part of the world. Again, we must be careful to distinguish in principle between genuine folk beliefs and the scholarly works which transmit these beliefs to us. On the whole it appears that the miners believed in two different kinds of mining spirit, the good and the bad. They believed that the bad spirits were of the devil and to be avoided at all costs. But the dwarfs were almost always benign. Georgius Agricola, writing in the sixteenth century and generally a sceptic in matters concerning the supernatural, described the kindly dwarfs in the following terms.

> Then there are the gentle kind which the Germans as well as the Greeks call cobalos, because they mimic men. They appear to laugh with glee and pretend to do much, but really do nothing. They are called little miners, because of their dwarfish stature, which is about two feet. They are venerable looking and are clothed like miners in a filleted garment with a leather apron about their loins. This kind does not often trouble the miners, but they idle about in the shafts and tunnels and really do nothing, although they pretend to be busy in all kinds of labour, sometimes digging ore, and sometimes putting into buckets that which has been dug. Sometimes they throw pebbles at the workmen, but they rarely injure them unless the workmen first ridicule or curse them. They are not very dissimilar to goblins, which occasionally appear to men when they go to or from their day's work, or when they attend their cattle. Because they generally appear benign to men, the Germans call

them guteli. Those called trulli, which take the form of women as well as men, actually enter the service of some people, especially the Suions. The mining gnomes are especially active in the workings where metal has already been found, or where there are hopes of discovering it, because of which they do not discourage the miners, but on the contrary stimulate them and cause them to labour more vigorously.

On this account the dwarfs do not really mine, though they pretend to. Nonetheless they are very closely associated with the metals, appearing only where the metals are already being worked or where they may be discovered. But the stories about dwarfs from this period go further than this, since they show how the presence of dwarfs was taken by miners as a sure sign that valuable ore was nearby. It appears that the dwarfs were regarded in some way as the spirits of the metals, so that where dwarfs were seen, the metals had to be also. Sometimes the dwarfs voluntarily showed the miners where to dig, and tales of this kind explained the discovery of some of the most valuable mining sites. Sometimes a miner succeeded in capturing a dwarf and refused to release him until he had revealed the location of a rich ore body. On other occasions dwarfs warned miners of dangers, as when the first man on the early morning shift raised the ore kibble and found a little man standing in it who pointed with a gesture of warning towards the shaft and then disappeared. When his fellows arrived the miner told them what had happened and while they argued about what it meant, the shaft caved in beneath their feet. Without the warning many of them would have been killed.

Then there is the famous story of the dwarf dance in the Copper Mountain, a valuable mine long since exhausted. There is a vault in this mine still called the dwarf chamber after the story which is told about it. Three miners once descended down the deepest shaft and, after working hard for most of their shift, were suddenly surprised by the most beautiful music which seemed to come from the interior of the mountain. This music was even better than the music played by the miners' band at the yearly festival of the miners' feast, and the three miners quietly prepared to depart so as not to disturb the spirit of the mountain. But as they did so they saw a great number of little men coming towards them, each hardly bigger than a human hand. The little men carried musical instruments, and after a little dance they greeted the miners and asked them to join them. They told the miners not to worry about their work, since they would themselves do anything left undone. The miners gladly agreed because they were tired. Then the dwarfs danced, jumping over each other and so fast that the whole mountain seemed to spin. The miners could not resist laughing at them but the dwarfs did not take it amiss. When the dwarfs had sat down again, one of them came up to the miners and touched their eyes. They felt they were blinded but the dwarf took them by the hand and led them to a chamber where they recovered their sight. The chamber was full of precious stones and gold and silver bars, stacked like pieces of wood in the kitchen at home. After a silence the dwarf told them to take what they wished, saying they would be happy provided they remained diligent and thrifty. When they emerged from the mine, the gold was still in their hands and each bought a little home and lived happily with his family. But later one of them became proud and thought he would not have to work anymore, and dire poverty overcame him because he had ignored the dwarf's warning.

In this story the dwarfs are both the keepers of treasure and miners. The treasure of gold, silver and precious stones is in a secret chamber of the mountain while the mountain itself is being mined for copper and iron. This poses no problem for the story teller, and neither does the dwarf's leading of the miners by the hand when the dwarf himself is no bigger than a hand. In the readiness of the dwarfs to complete the miners' unfinished work we see again how these spirits or sprites of the metals aid human beings in their mining endeavours. From the dwarfs' point of view the mining by humans of their underground homes is entirely proper. This is further emphasised by their insistence that the miners continue to live diligent and thrifty lives despite their new wealth. Dwarfs are in some sense spirits of nature, so we may infer from these stories that human mining has its place in the natural order of things. The description of the gold and silver bars stacked like firewood in the kitchen at home is especially charming. The dwarfs can afford to be a little careless of their wealth when they have so much of it. For their part the miners are scrupulous in their dealings with the dwarfs, prepare to quit the mountain even though they find the music wonderful, and are anxious lest the dwarfs take their laughter at the dance amiss.

The dwarfs are a class of elves or nature spirits who live underground and are associated with the metals. In order to understand how they come to be as they are in folklore, we must consider the larger scheme of nature spirits among whom they take their place. These belong to a conception of the world which predates Christianity and may even predate the Olympian religion of archaic and classical Greece. In a famous poem the seventeenth century poet John Milton explained the conception, dating it back to Plato and to Hermes Trismegistus, the thrice great Hermes who was identified with the Egyptian god Thoth.

> Or let my lamp at midnight hour
> Be seen in some high lonely tower,
> Where I may oft outwatch the Bear

With thrice great Hermes, or unsphere
The spirit of Plato, to unfold
What worlds or what vast regions hold
The immortal mind, that hath forsook
Her mansion in this fleshly nook:
And of those demons that are found
In fire, air, flood, or under ground,
Whose power hath a true consent
With planet, or with element.

In these lines Milton imagines himself studying the work of Hermes till all hours. He would evoke the spirit of Plato to reveal to him the destiny of the immortal mind after it has left the body, and to explain the natures of the elemental spirits. It is not clear from the passage what connection Milton perceived between the discarnate mind and the elementals, but he certainly believed that Plato could explain both to him. He calls the elementals demons and we must be careful to distinguish here between pagan and Christian demons, given that Milton is explicitly concerned with pagan traditions. In classical Greek usage the word demon is used for divine and semidivine beings without any connotation of evil or devilishness. In Milton's lines they are the spirits of the four elements. These demons all have a certain power which is unspecified but which has a true consent with planet or with element. Why did Milton and his midnight teachers suppose that the elements were invested with demonic power?

According to the writings of Hermes and Plato the creator of the universe fashioned first the stars and planets to which he delegated the authority for organizing the rest of creation. According to Plato the creator also devised the elements, but the making of the other kinds and species of creatures devolved upon the astral and planetary powers. These astral and planetary powers were supremely intelligent as was shown by the predictability of their movements. The more perfect the creature, the more regular its movements. On the one hand they were turned towards the creator whose divine plan for the world they contemplated; on the other they were turned towards mortal creatures as they strove to realise this divine plan in the world. According to the Hermetic writings, the lesser gods, and especially the sun, achieved their work of creation by means of genii whose task it was to fashion the material of the world into the various forms which comprised the divine plan. These genii were exceedingly numerous. In the human constitution for example, they imprinted their likenesses on souls, and were present in the nerves, marrow, veins, arteries, brains and viscera. In short every creature, animate or inanimate, owed its determinate form to these genii who worked constantly to maintain it. From this point of view everything that happened in the world, from the catastrophes which overthrew nations to the smallest involuntary movements of the human body, was conducted by spirits.

The genii or demons were the links between the astral powers and the kinds and species on earth. They were the means by which the stars and planets organised the rest of the creation. In this way all the kinds and species were affiliated with the stars. Every plant, for example, was patterned after one or other of the astral bodies and all the plants together realised at their level the entire scheme of the heavens. The same was true of birds and beasts, cities and nations, and the different human types and characters. In the case of the metals these links with the heavens were felt to be especially strong, and the names and signs of the planets were often used as the names and signs of the different metals according to the following table.

Saturn	Jupiter	Mars	Sun	Venus	Mercury	Moon
Lead	Tin	Iron	Gold	Copper	Mercury	Silver

The relation between a planet and a metal was much more than an abstract correspondence. According to many ancient, medieval and renaissance thinkers, the planets were the efficient causes of the metals in the earth. Their influences actually brought about the existence of the metals. This was the function of the genii or demons who emanated from the celestial sphere, entered into the earth and fashioned therein the various ores and veins. In the case of plants and animals it is not difficult to envisage how earlier times could have conceived of them as animated by nature spirits. It is rather more difficult to conceive of Naiads and Oreads, the nymphs of streams and mountains. With the metals, however, we have somehow to suppose that dwarfs were the intelligent, formative powers within them, the means by which the metals came to be in the earth, the agents of the astral powers at this level of the creation. The dwarfs were at once the makers, owners, keepers, miners, smelters and workers of the metals because they were the occult intelligence of the metals according to a system of thought which animated the entire creation. In one respect the dwarfs and metals were at the outermost limit of this system, hidden in the depths of the earth. But, as is often the case, the lowest rank is very close to the highest in its own special way. The metals and their informing spirits were among the clearest examples of the planetary powers in nature. In this way the whole of the creation was most intimately bound together.

One of the greatest and best known stories about mining dwarfs is the story of Snow White. There are many versions of it. One of the earliest is given by the brothers Grimm who copied it down from the words of an old woman at the beginning of the nineteenth century. Let us assume that Snow White symbolises the soul or psyche and that the whole story is concerned with the soul's journey through life. It is worth comparing the story of Snow White with another great fable of the soul's journey through life, the fable of Psyche herself as told by Apuleius in the Golden Ass. Like Psyche, Snow White is the daughter of a king and is born in a palace, which we may take to represent her heavenly station in the spirit. Similarly her marriage to the prince at the end of the story signifies her return to her former station, her resurrection in the spirit after succumbing to the poison of the apple. In answer to her mother's prayer, she is as white as snow, as red as blood and as black as ebony. These three colours, as we have seen in alchemical theory, represent three fundamental tendencies or potentialities within the soul. Her ejection from the palace of her father by her wicked stepmother represents the soul's fall from its divine station. If we ask at this point how the divine can be compromised in this way if it is truly divine, we may answer that the reason for the soul's fall from its divine station is a mystery, and that this mystery is often represented, though not explained, by the force of envy. The wicked stepmother wants to be the fairest in the land and tries to destroy Snow White to rid herself of a rival. Similarly in the story of Psyche, Venus is envious of Psyche's beauty and the honours paid to her, and so arranges her downfall.

After being turned out of the palace by her stepmother, and having persuaded the huntsman not to kill her, Snow White makes her way over the seven mountains to the cottage of the seven dwarfs. These dwarfs are miners who dig and delve in the mountains for ore. Snow White is impressed by the neatness and cleanliness of the little cottage which the dwarfs maintain with a truly Germanic sense of order. Since Snow White is still only a child, the furniture and utensils of the dwarfs are the right size for her. Some of the beds, indeed, are too big for her. She eats something from each plate, drinks something from each cup and lies on each bed until she finds one that suits her. There she says a prayer and goes to sleep. When the dwarfs return they light seven candles, discover that the perfect order in which they left the cottage has been disturbed, and finally they find the sleeping girl, around whom they stand with their seven little candles. Then the dwarf whose bed she had taken takes turns in sleeping in the beds of the other dwarfs, an hour in each. When Snow White awakes in the morning, she tells them her story and they ask her to stay with them, promising to provide her with everything she needs. In return they ask her to serve them, specifying seven duties which they require her to undertake.

Snow White and the dwarfs have much in common. They are all, we feel, exiles. Snow White has been cast out and nearly murdered by her step mother. She has run as far away from the palace as she could, over the seven mountains, and there she has come upon the cottage of the dwarfs. They too are exiled, living far apart from the rest of humankind, partly because their vocation as miners requires of them that they work the desolate places of the earth, and partly too, we feel, because their dwarfishness has marked them off from other people. When we read of the seven mountains and the seven dwarfs who work them, it is hard not to think of the seven metals even though the dwarfs mine only copper and gold. Dwarfs, as we have seen, are the intelligent genii of the metals, the agents of the planetary powers under the ground. If Snow White's exile is the soul's fall from its divine station, then her residing with the dwarfs symbolises the soul's involvement with the planets when it enters into a mortal life. The seven dwarfs and the seven mountains are symbols of the planets which govern and guide all the creatures on earth.

This is why Snow White eats from each plate, drinks from each cup and lies on each bed. Every human soul born on earth acquires its faculties and powers from the seven planets as it passes through their orbits on its way to earth. This is the meaning of the tableau in which the seven dwarfs stand round the sleeping girl with the light of their seven candles shining upon her. And this too is why Snow White is given seven tasks to perform in return for her keep, because the various faculties and powers of the human constitution are bestowed upon us by the planets, just as the gods and goddesses of Olympus bestowed the gifts of the arts and crafts on the golden maidservants of Hephaestus. But for all its neatness the cottage of the dwarfs is still only a cottage; it is not a palace like the one from which Snow White has come and the one to which she is destined to go. For all their goodness, the dwarfs cannot save her from her stepmother. They are little people who work underground, and Snow White's time with them symbolises the soul's exile from heaven, a mortal life. From the same point of view Snow White's sojourn with the seven dwarfs symbolises her subjection to time, to the seven days of the week. She is strictly bound to their routine as housekeeper.

Snow White's stay with the dwarfs is interrupted by three visits from her stepmother. On the first she is asphyxiated by the laces of her bodice and falls down as if dead. This attack upon her corresponds to her whiteness since it is her chest which is attacked. On the second visit she is poisoned by a comb in her hair and falls down as if dead. This attack corresponds to her blackness. On the third visit she is poisoned by the red part of an apple, an apple which is part wholesome and part deadly. This time she really falls down dead and the returning dwarfs are unable to revive her.

The redness of the apple corresponds to her redness and it is through her lips that the poison enters. That eating should precipitate her second fall has many parallels. Adam and Eve become mortal through the eating of an apple which is also part wholesome and part deadly, from the tree of the knowledge of good and evil. Persephone is committed to the god of the dead when she tastes the pomegranate seed. But even though Snow White is dead she does not decay, and so the dwarfs put her in a glass coffin which they guard but do not inter. Her corporeal preservation is a symbol of the soul's indestructibility and her being guarded by each of the dwarfs in turn shows that she is still under the protection of the planets and subject to time.

With the arrival of the prince, her resurrection begins. It is worth comparing the latter part of this story to the alchemical process. When she falls down dead from the apple she enters the black stage. When she is placed in the glass coffin she is in the white stage, the state of pure passivity and potentiality. With her marriage to the prince, the solar hero, the alchemical marriage is accomplished, the sulphur and quicksilver perfectly combined, the soul and spirit united. Snow White becomes the mistress of her own palace, and the wicked stepmother who brought about her double fall is done to death by the wedding dance. The old order gives way to the new, and everything which had hindered the process of development is found at last to have helped bring it about. Like the dragon Holdfast, the stepmother had striven to keep things as they were. It is entirely apt that she be destroyed by the dance which inaugurates the new world of the soul's fulfilment.

This story makes the fairy miners the central symbol of the human condition, and implies the intimate connection between the planets and the metals which we have already considered. The dwarfs symbolise the elemental powers of the human soul which they may protect but which they cannot liberate. Nonetheless they give up Snow White to the prince as a gift and, unlike her stepmother, are only too happy that she should come into her own. They give her up as a gift and would refuse all the gold in the world in return for her. In this way they show that they value the soul's redemption more than all the treasure their mining could win. Yet for all their power the dwarfs are subordinate to the great battle of good and evil which is fought over Snow White, and this may give us the clue to their size. In the Christian world the fairy spirits of nature belonged to an order of belief other than that of the Trinity or the saints. Being of the natural world, they were severely circumscribed in their exercise of power since they did not fully belong in the realm of the spirit. That Snow White should say a prayer before she sleeps in the cottage of the dwarfs indicates that she is in some way beyond them. Their size and the size of the other elves may reflect this ranking of powers, just as the lameness of Hephaestus may have reflected his involvement with the material world.

Chapter Six

THE DESACRALISATION OF WORK

And if a man takes upon himself in all its fullness the proper office of his own vocation, it comes about that he and the world are the means of right order to each other... For since the world is God's handwork, he who maintains and heightens its beauty by his tendance is cooperating with the will of God, when by the aid of his bodily strength, and by his work and his administration, he makes things assume that shape and aspect which God's purpose has designed. What is the reward?... That when our term of service is ended, when we are divested of our guardianship of the material world and freed from the bonds of mortality, God will restore us, cleansed and sanctified, to the primal condition of that higher part of us which is divine.

Hermes

FROM THE POINT OF view of the earlier chapters of this book, the current debate about mining is exceedingly superficial. All parties to the current debate ignore and are almost certainly unaware of the profound spiritual history which transforms the theory and practice of mining for those who know it. Ignorance of this spiritual history is typical of a time in which people have forgotten that work is holy, and is none the less holy even when its holiness is no longer acknowledged. No one can understand the debate over mining except in this larger context of its sacred history. The story of how work was successively desacralised in the western world from the time of the renaissance, and the light which this story casts on mining are the subjects of this chapter. The telling of this story will take up many of the theories which have been considered in earlier chapters.

THE TRADITIONAL WORK ETHIC

In the west, as we have seen, the earliest theorist of work at any length or in any detail was the Greek philosopher Plato. His account of work has remained one of the standard accounts. An almost identical theory to Plato's is to be found in other cultures and particularly in India where it underlies the practice of Karma Yoga, the yoga of action. This theory, wherever it is found, may be called the traditional or perennial theory of work, and most of what has been said in the last four chapters derives from it. A brief outline of its major tenets should therefore be enough to establish it clearly and firmly in the mind of anyone who has read this far. But we should take care from the beginning not to confuse this traditional theory of work with a theory which superficially resembles it, the theory of work which is usually called the Protestant work ethic. Though both of these theories suppose a relationship between work and the life of the spirit, they do so in quite different ways. Any understanding of the traditional theory of work must also be an understanding of how it differs from the Protestant work ethic which largely displaced it.

According to Plato everyone born into this world has an innate predisposition for a particular kind of work. Only by the finding and doing of this work can a person become who he or she truly is. This predisposition is the single determining factor of the human personality, in comparison to which all other traits of character, accidents of birth, environmental conditionings are negligible. Each of us is born to carry out a particular task and only when that task is completed have we done what we came for. In some people this predisposition is very clear, as in the case of child prodigies who evince at an early age a degree of competence in a particular art or science which is inexplicable in the light of their actual experience. According to the Hindus this is one of the strongest reasons for a belief in some form of metempsychosis or transmigration of souls from one body to another. So natural and unforced is the facility which a predisposition confers that the person so gifted is hardly aware of it. It is only with difficulty that a child who can draw can be made to understand that others cannot. There is some evidence to suggest that these predispositions run in families, but both Plato and the Indian philosophers are careful to point out that there is no guarantee of this, and that in a well organised society people are free to pursue other vocations than those of their parents. Nonetheless there is an expectation in traditional societies that children will follow their parents in this regard, and this expectation, taken together with the central importance of these predispositions to the personality, explains why families are often named after vocations. In north western Europe the names of Bergman and Smith are particularly common and derive from the professions of mining and metallurgy. Sworder is another such name but much less common.

To a society like ours which has largely done away with the traditional arts and crafts, it may appear that they are the products of convention rather than of nature, and that they can be dispensed with when cheaper and more efficient means of production are discovered. This assumption is

122

debatable. One of the most remarkable developments of the two centuries since the industrial revolution is the hobby. After working in the factory or the office people return home to practice in their periods of leisure what previously they would have done as work. This is the significance of gardening in a society which has mostly dispensed with agricultural labour, and of the millions of workshops in the backyards of suburban houses. Nothing could show more clearly than this that the old predispositions continue to exercise their sway over the personality, and they do so regardless of the fact that the work for which they fit us is no longer paid, nor otherwise rewarded than by the intrinsic satisfaction which it provides. When Plato starts to talk about work in the *Republic* this is the very first point he makes. He asks whether people would be better off if each did or made everything, or whether each should do or make one kind of thing only and then share the fruits of this labour with everyone else. In deciding that it is better to divide labour than have each person do everything, Plato argues that each person is naturally fitted for one kind of work only and is better served by doing just that. For Plato the prime reason for dividing labour is not that it is more efficient, but that it conforms to our innate predispositions. Acting in accordance with one's innate predisposition is the basis of Plato's theory of justice.

In the Indian philosophy this same theory or law of justice is called the Dharma and it is one of the major themes in the best known of all Indian scriptures, the *Bhagavad Gita*. As the Indians understand it, we are impelled into action by the mere fact of our bodies. We cannot do other than act, given our equipment of arms and legs. Actions are constantly flowing from us, and what is required if we are to be happy is a way of organising and directing this ceaseless flow of actions to some worthy end. This will finally enable us to free ourselves from the otherwise unending chain of causes and effects by which our actions bind us. This release is achieved by the selfless performance of our proper work, without any regard for the fruits of it, until we become capable at last of a kind of desireless, actionless action which is liberation. This way of thinking has something in common with the story of the fall from the garden of Eden. Before the fall Adam tended the garden, cooperating in the work of God. But after eating the fruit of the tree of the knowledge of good and evil, he was expelled from the garden and forced to provide for himself by the sweat of his brow. This concern with good and evil is precisely what transforms the selfless work of the one who is liberated into the anxious toil of the fallen. As the *Bhagavad Gita* puts it:

> Without hope, with the mind and self controlled, having abandoned
> all possessions, doing mere bodily action, he incurs no sin.

> Content with what comes to him without effort, free from the pairs
> of opposites and envy, even minded in success and failure, though
> acting he is not bound.

This is how the *Bhagavad Gita* describes the man liberated through work. This selflessness in action is characteristic of the work ethic in traditional societies. The work is done anonymously and workers do not seek to arrogate to themselves the credit for having done it. This is why much of the greatest work done in the middle ages, or in the archaic period of ancient Greece, is unsigned and unattributed. The practice of claiming work as one's own is an index of the extent to which the traditional work ethic is in decline, and on this score both the classical period of Greek art and the renaissance are in the process of falling away from the selfless ideal of the periods which immediately preceded them. In the ancient world this falling away was merely a matter of degree, since in the Greek and Roman traditions as we have seen, credit for the work had always finally to be given to the patron god or goddess. It was always believed that a divine power was responsible for the miracle of skilful action or creation.

These ways of thinking about work made it clear that workers were not to use their work to aggrandise themselves, at whatever level in the society they might be. Nor was work, even of the most artistic kind, a medium of self expression, in which the personality of the artist as an individual was exposed. The personalities of artists were of no more interest to those who made use of their work than the personalities of their tailors or cobblers. At best they had nothing at all to do with the work. In such a society each kind of work was done by people who sought through the doing of their work to escape the limitations of the egoic self. Instead of thinking of themselves as individuals who were as far as possible separate and independent of each other, they thought of themselves as belonging to parts of a single organism. These parts were the classes and professions, each of which was different from the others, but necessary to the survival and success of the whole. Just as the same food produces and maintains the different organs of the human body, all of which are necessary to its fulfilment, so in the one society all those innate predispositions were to be found which were needed to complete it. Acting in accordance with one's innate predisposition was justice which was at once the source of the deepest satisfaction to oneself and the means of maintaining the society.

In many different ways the social order was continuous with the natural order. The innate predispositions which equipped people for particular kinds of work were in nature in much the same way that we now consider, say, the home building instincts of animals to be. The idea of whatever was to be done or made stood in the divine mind in exactly the same way as did the ideas of the natural species or those of the elements. The contemplation of the idea was the superior part of the different kinds of work, while the material realisation of the idea in the world of time and space was regarded as derivative and secondary. Since every such idea came from God, it could hardly be regarded as the creation of an individual craftworker or artist, and therefore the notion of originality counted for very little. This is not to say that research, experiment and innovation were suppressed. Plato was emphatic that enquiry is essential to the proper development and maintenance of any art or science. Instead there was a tendency to attribute the latest finding to some earlier, often legendary exponent of the art, as a token of veneration and as a way of ensuring the continuity of the tradition. In some places the same way of doing or making things persisted for many centuries, as was the case, for example, with Gregorian chant. But it would be wrong to suppose that the artists at the end of one of these periods were less capable than those at its beginning. For traditional workers, originality consisted in the recreation within themselves of that understanding which the centuries had inherited from the founder of their art, who had received it from God.

Another respect in which human work was continuous with the natural order was in the relation between the worker and the material on which the work was done. Of almost all kinds of work there was an assumption that between the worker and the material a bond existed, a deep affinity. The carpenter had a feel for the wood, the smith for the gold or iron, the gardener had green fingers. This affinity, which underlay the activity of most working lives, established a connection between the deepest element in the personality of the worker and the universe beyond, between the microcosm and the macrocosm. This was no abstract speculation but an immediate recognition that by working through the creations of the outer world of nature, a vocation could be answered and a life fulfilled. This connection between the innermost and the outermost dimensions of experience has much more to do with human happiness than is now realised. It is the only means to the thorough integration of the human being, and the loss of it produces an alienation far more pervasive and acute than that described by Marx. Between the idea which is known through contemplation, and the material through which that idea is realised in the world of time and space, there may be a union which is the marriage of heaven and earth.

Just as the worker required the material on which the work was done in order to achieve fulfilment, so the material required the worker if it too was to be fulfilled. It seems strange to us to suppose that wood can only be fulfilled, come fully into its own, through the intervention of the carpenter. On this view the wood's fulfilment consists not in its living out its full span in the forest, but in being axed and sawn, sanded and polished. Only then are its beauties revealed. A cathedral mason who ruined the block of stone he was dressing was required to follow the cart which disposed of it as chief mourner. As stone the block would not be affected much by whatever the mason did to it. But by being turned into a block it became, as it were, alive, and then died again through the incompetence of the mason. For many traditions the metals in the ground were embryonic and the processes of mining and smelting were obstetric, bringing them to birth. According to some scholars this is the primordial view of mining, the oldest and most profound. But what all these examples have in common is the notion that nature longs for human intervention in order to become most fully itself. Intervention by humans in the natural order is not a rape but nature's glory, the only means by which the greatest treasures can be brought to light. As William Blake put it:

Where man is not, nature is barren.

The alchemists put the same thought even more abruptly:

Nature unaided fails.

It is hard for us now, two hundred years on from Wordsworth, to realise that for by far the greater part of recorded history, wilderness was not regarded as beautiful but as ugly and frightening. Very often it was believed demonic, in the bad sense, a natural condition quite different from that of Eden which was a deliberately planted and tended garden. For the greater part of Christian history, wilderness was precisely the natural correlative of Adam's fall, a place of no virtue except for what could be won from it by human effort. There was normally only one class of people attracted to it for its own sake, the anchorites and hermits, for whom it was at once a solitude and a test, a retreat from the world and an arena for spiritual combat. Even so, this use of the wilderness is far from universal, appearing most often at times when the social order is markedly decadent. At those times the most extreme luxury and the most extreme asceticism are found side by side, and from the point of view of several religious traditions the luxury and the asceticism are equally suspect. It is one thing to retreat to an ashram in the forest and quite another to endure the terrible privations of the desert. The place presently occupied by the notion of wilderness in our range of emotions and attitudes was in early times taken by pasturage and the pastoral tradition. This probably means that we can no longer quite grasp what the idea of wilderness was for our ancestors. For us there are no longer wild and dangerous places on the earth as once there were. For us a wilderness may also be a park, a confusion of categories impossible before.

But if the wilderness was almost entirely bad from the point of view of earlier times, so also were those who did nothing to transform the raw materials of nature, of whatever kind. People who failed to respond to their vocations, who did nothing to develop themselves in accordance with their innate predispositions, were despised and condemned. On this view everyone is born to be an artist in some field; there is nothing finally to distinguish the work of the person whom we now call an artist from the work of anyone else. The artist, as one scholar has put it, is not a special kind of person, but everyone who is not an artist in some field, everyone who does not respond to their vocation, is an idler. The traditional theory of work cannot, I think, conceive even of the possibility of a person without an inborn vocation, since the vocation is the single greatest determinant of personality. There is only one person who has the right to abstain from all constructive activities, the contemplative monk or nun. But these people not only make nothing, they do not use anything either. In a strict sense they are no longer members of human society. But even these people may be said to engage in the work of transforming the raw materials of nature, since they are wholly intent on correcting and improving their own fallen natures as human beings. And, of course, it is the activity of these people which we still think to be the most truly vocational of all.

For all these reasons sloth was one of the seven deadly sins in the medieval understanding. But the traditional doctrine of work entailed much more than the mere banning of idleness. Other activities were regarded with the deepest suspicion. Merchants, for example, were typically unproductive. Engaging in no constructive activity of their own, merchants bought cheap and sold dear, and it was not easy to find a justification for this within the limits of the doctrine just outlined. This problem was, however, peculiar to the West; in India the merchant class was always highly respected. Still harder to accommodate, but also apparently indispensable, were the money lenders, around whose operations the later middle ages built an enormous scaffolding of casuistical argument and counter argument. The traditional argument was that metal, the metal of the coinage or the precious metals, was essentially barren and could not in the normal course of nature propagate itself. But this is precisely what it did when lent out at interest, and therefore the lending of money at interest was unnatural, a sin against nature. Though not a deadly sin in itself, it was very closely connected in the medieval mind with sloth and avarice, so that throughout this period the merchant and money lender alike were thought to stand in imminent danger of hell.

These consequences of the traditional doctrine of work were brought out very clearly in a passage of Dante's *Inferno* in which Virgil explained to Dante the reason why usurers were condemned by divine justice to suffer the torments of hell:

"Would you go back a little way," I said;
"To where you said that usury offends
Divine goodness, and untie that knot for me?"
"Philosophy," he said, "to him who heeds
Indicates in more than a single place
How Nature derives the course she follows
From the divine intelligence and skill,
And if you study the 'Physics' carefully,
You will find, after not too many pages,
That your human skill, as far as possible,
Follows her, as pupil follows teacher,
So that your skill is like a grandchild of God.
From these two, if you recollect
The opening part of Genesis, mankind
Must draw its sustenance and move ahead.
And because the usurer pursues another course,
He scorns Nature for herself and for her
Follower, and sets his hopes on something else."

Virgil's source for some of this is the *Physics* of Aristotle, though he goes far beyond what Aristotle says there. All human activity, on this view, should model itself on nature, which models itself on the divine intellect. In this way, by depending on art and nature, humankind should gain its livelihood and develop. Human art repeats and imitates the creative powers of nature on the one hand, and the creator of nature itself, the divine intellect, on the other. As for Plato, so for Dante there is something divine about human work. On Dante's account the usurers' sin is one of omission. They produce nothing and in that failure are guilty of despising God and nature. They are guilty of violence towards God, and their punishment is to be condemned to the burning sands of a desert plain where fiery embers fall on them like flakes of snow.

Virgil explains to Dante that human work is an imitation or repetition of the divine intellect. It is just here that we see the significance of Homer's accounts of Hephaestus and of the revelation to Moses on Mount Sinai. These stories are exemplary. They do not apply merely to the making of Achilles' shield or the vessels of the tabernacle. They show that the work of the smith is always a repetition of the divine act of creation. Every craftworker realises in human form the creative power of God, just as every contemplative realises the divine inactivity and inwardness of God. The whole world of human work is a bodying forth at many levels of the different aspects of the divine nature. By a splendid anachronism Dante's Virgil makes his point about God, nature and art by referring both to Aristotle and to Genesis.

THE PROTESTANT WORK ETHIC

Dante composed the *Inferno* about 1300 AD. From that time to the time of William Blake was a period of five hundred years, during which the traditional doctrine of work was successively forsaken until it was almost forgotten. If that should seem a very extended period for the demise of a single theory, we must remember that this theory was the basis for much of the psychology, sociology, economics and politics of the traditional order, and that it had persisted in the west from at least the time of Homer. There are reasons for believing that the theory played a central part in Egyptian civilisation from early times, which would date it back another two millennia. It is not surprising therefore that a theory of this age and importance took a long time to disappear. We must now study its decay in some detail since it is precisely from this decay and its consequences for our understanding of work and nature that the terms of the present debates about mining have emerged.

At some points it is easy enough to trace the slow decline of the traditional doctrine of work, especially in the later stages. This is because the decline corresponded to the emergence of a new doctrine of work which has been of the greatest interest to historians and sociologists. This new doctrine is usually called the Protestant work ethic, though this name is not entirely satisfactory. This new ethic is generally thought to be the religious and psychological basis for the rise of capitalism in the western world from the sixteenth century onwards. It has been minutely studied in an attempt to isolate those factors which created the capitalist order. Of all the questions and issues raised by historians and social scientists since the beginning of the nineteenth century, this is one of the most vital.

Unfortunately from our point of view, this great question has been discussed almost entirely the wrong way round. The task has not been to explain how the older theory of work declined but how the new one emerged. There has been a presumption, at least from the time of Marx, that the formation of capitalism is a positive development, a social and economic order into which Europe grew, a stage on its journey towards full maturation. Accordingly, very little attention has been given to the virtues and values of the order which preceded capitalism and the new ethic of work. The best known historian of the new ethic was Max Weber who claimed that Luther originated the notion of vocation, for which there had been no historical precedent! This perhaps is an exceptional case, but there can be no doubt that students of this period have given very little weight to the theory of work which the Puritan and capitalist revolutions deposed. To trace, therefore, the demise of the older doctrine is to take on a task that has been little attempted.

The decline of the traditional theory of work was bound up with the decline of the middle ages. The feudal order of the middle ages was irreparably damaged by the Black Death, which created so great a shortage of agricultural labourers that the manors could no longer enforce the service of their serfs. Their labour was at a premium and could be sold to the highest bidder. The weakening of the manorial system encouraged an increasing lawlessness in the countryside and a shift to the towns. To sustain the towns, new manufacturing industries had to be developed which produced goods for export and not merely for local consumption. The most important of these in the early period were woollen goods and the towns which produced the best woollen products were in the Lowlands. These towns developed their woollen manufactures to the point where they needed to import raw wool from further and further afield, and so there began the conversion of agricultural land into pasturage in many parts of Europe. This in turn led to a still greater exodus from the land, so that by the early sixteenth century Thomas More could describe the sheep as a man eating animal since it deprived agricultural workers of their livelihood.

As the power of the towns increased, the power of the landed nobility declined, and money as well as land became a standard of wealth. This transference of power was a serious, if invisible, blow to the Church which had come to mirror the world of the peasantry and the countryside in its calendar and rituals. The rich townsfolk did not owe their wealth and power to a long established order, and from the beginning of the renaissance a new emphasis on the individual appeared. The patronage of the arts was also in part transferred, as the new wealth gave the direction of the crafts into the hands of the merchant princes. This was particularly true of Florence, the greatest woollen manufacturing city in Italy, but it can be seen also in the paintings of the Lowlands. Personal portraiture reemerged on a large scale after its eclipse during the middle ages, as the merchants redirected the arts to the task of immortalising themselves rather than of glorifying God. The artists, too, forsook their personal anonymity and quickly developed a remarkable bravado and braggadoccio, as can be seen, for example, in the autobiography of Benvenuto Cellini.

This increase in trade between cities and nations required a corresponding development in the mechanisms of international finance. The first great private bankers had acted as agents for the Church in Rome, to which they facilitated the payment of tithes and taxes by means of letters of credit. The fact that such methods were used by the Church itself made it much more difficult to maintain the religious strictures on merchants and usury. One great historian of usury during this period, R.H. Tawney, traces in detail the slow relaxing of the rules concerning usury under the pressure of the new economic order. According to Tawney, the accumulation of wealth during the late middle ages reached a critical point in the fourteenth and fifteenth centuries. At this point it burst the narrow limits within which it had grown to that time, and then reconstituted the social and religious orders to accommodate itself.

The development of trade and international finance weakened the nexus between people and the locale in which they lived. The idea of self sufficiency as a desirable goal for a society gave way to an increasing appetite for the exotic. The limited aims of medieval government yielded to quite new ways of calculating political and economic success, of which the possibilities were continually being enlarged. As more and more people encountered products from abroad, the connection between the immediate environment and the amenities and utensils of daily life was lost, and this brought about a revision of attitudes to the natural world. To this change the discoveries of the new world were soon to make a massive contribution, not only in the natural sciences but because the wealth which flowed into Europe from the other side of the world had no immediate relation to the places which it enriched. The marriage between heaven and earth which the crafts had achieved through their transformation of natural resources immediately to hand, became less and less central to the economic arrangements of society.

It is an old belief that social institutions can be destroyed only by the corruption of those who govern them. In 1510 Martin Luther went to Rome and was appalled by the decadence which he saw there, particularly the sale of papal pardons for mortal sins. The same reforming anger which inspired Luther to attack the system of indulgences drove him on to attack other traditional accretions to the word of God in the Bible. According to Luther pilgrimages and saints' days were an excuse for idleness and should be done away with. He had a profound distrust of mendicant friars and the contemplative orders. He was a strong believer in the value of work and vocation, but these beliefs had much more to do with their moral than their spiritual value. Work had to be done because it was given by God and because it served the community. But it was not itself a spiritual path. At best for Luther it was a means of keeping the soul from the temptations of leisure and wealth. Hard work was good for the soul as a duty and a discipline, but finally it had nothing to do with salvation. Luther's hatred for the papal traffic in salvation led, from the traditional point of view, to a dangerous overreaction. Anxious as he was to ensure that salvation ceased to be a marketable commodity, he detached it from the social world altogether and made it entirely a matter of faith.

At the first level Luther's distrust of idleness made him emphasise the value of work more than had his predecessors. We can already see in Luther that tendency to think of work as a mechanical discipline which was to play a central part in the industrial revolution. But from the traditional point of view Luther's notion of work was very limited. He was exclusively concerned with the physical act of labour in this world, the slavish element of work. We do not hear from Luther, nor indeed from any of the reformers, about the contemplative or free act which must precede and accompany the realisation of the idea so contemplated in the world of time and space. We may put it simply by saying that they had not fully understood the exemplary nature of those metallurgical revelations which Moses had been shown on Mount Sinai. They underestimated or ignored the element of contemplation in the practice of the arts and crafts, paying little attention to the question of where those ideas originated of which every work of art or craft is an embodiment. Of a piece with this oversight was their destruction of the mendicant and contemplative orders, on the ground that these people did no productive work. Far from believing that contemplatives were the highest class of humanity, they refused to acknowledge them at all. They supposed instead that everyone should commune with God in the same way, without intermediaries, in the solitude of prayer. It is as though suddenly everyone was required to be a monk or a nun while still living in society. It may be claimed of this as of other forms of inflation, that it brought an apparent increase in wealth, followed by a long impoverishment.

Luther's view of work, with different emphases and some important modifications, was the view shared by almost all the reformers of the next two centuries. For Calvin, for example, work had even less to do with salvation than it had for Luther. Calvin's doctrine of predestination made salvation a gift of God, given irrevocably and without any consideration of the virtue or otherwise of the soul to which it was given. By just so much did Calvin suppose the gift of salvation to be above the human capacity to earn or deserve it. The moral life for Calvin was not a means to this immeasurable grace but a sign of it. The soul which was saved had only one goal in the world, to glorify God in every word and deed. This theory of the moral life and of work is not unlike the traditional doctrine in one respect, since there too the worker works for no advantage extrinsic to the work, but achieves a divine selflessness through his commitment to the work for its own sake. In the traditional view the fully realised worker is concerned only for the good of the work to be done. But in the traditional view the doing of the work is itself the means of achieving this selflessness, which is not predestined but won. And this selflessness is found at the very heart of the work, where the innate predisposition of the worker for that particular kind of work turns out to be a divine genius which transcends the limits of the human, but is at the same time peculiar to that kind of work and no other.

Calvin's theory of work was more like the traditional doctrine than was Luther's in this respect but it was less like it in others. In social theory Luther was conservative, with a medieval belief in the life of the peasant and a medieval distrust of trade. For Luther the vocation to a traditional kind of work was of vital significance, especially in his later writings, because it was the principal means of service to the community. The smith's work was priestly because it served the community. Luther was himself the son of a miner. In Calvin's thought the traditional vocations were by no means central. His social theory was less peasant than urban, and applied as well if not better to the life of trade as to the lives of craft and field. He was concerned with questions of fair exchange and fair profit and was prepared to engage in the complex economic considerations that these issues demanded. Though his religious vision enabled him to create a social organisation at every level and on a grand scale, that organisation was in theory if not in fact very different from the social system of the middle ages. Gone is that governing ideal of integration in which all the different functions of the social organism were predisposed by divine providence to meet the needs of the whole and to fulfil the differing talents of each. In its place was a social order derived from and consistent with the doctrine of individual predestination, a city of the elect marked off from the rest of humanity and held together by a communal sense of this unbridgeable gulf.

In their own lifetimes the teachings of Calvin and Luther profoundly altered the societies in which they lived. They were men whose time had come, and continental Europe was instantly changed by them. Their effect on England was less immediate but when it came it was even more profound. In the meanwhile, throughout the fourteenth, fifteenth and early sixteenth centuries, there were other tendencies in English life which ran counter to the general movement from the feudal to the mercantilist order, or at least counter to the desacralisation of work. During these centuries there was an increase in the number and power of the craft guilds. These guilds conferred upon their members a strong sense of how their work was valuable of itself. Each guild was proud of the peculiar nature of its work, the special knowledge which its members shared to the exclusion of outsiders. Quite new occupations quickly developed this sense of their own mystique so that very soon after their invention printing and gun making were organised on the basis of guilds.

Each of the crafts had its patron saint and its own shrine, the maintenance of which was the responsibility of the guild. The members of the guild prided themselves on their exclusive privilege of worshipping at that shrine. Each craft had its own holy days in honour of its patron saint and took part as a guild in other festivals of the Church. In many parts of England the guilds were charged with the responsibility for putting on the miracle plays, the great cycle of plays which the Church mounted annually for the edification and education of the townspeople. In the city of Chester each of twenty five such companies put on a play. Each of the guilds had its own uniform or livery, to be worn by its members on special occasions.

This practice and others are still preserved by the livery companies of the City of London. There were also the apprenticeship rituals which further developed the craftworkers' pride in the mystique of their work. We know from many sources that the apprentices of each craft formed a very tightly knit body which often made riot in the streets with cudgels and barricades against its rivals. Each group had its own oaths and boasts. The good natured rowdiness of the shoemakers is dramatised in Thomas Dekker's play 'The Shoemakers' Holiday' which was published in 1600. The play pictures the city life of Elizabethan England though it is nominally set in the reign of Henry V. It tells the story of how the yearly feast of apprentices was established by Simon Eyre, shoemaker and lord mayor of London. It was a matter of the greatest pride to shoemakers that they should be the hosts on this occasion to all the apprentices of London. Dekker's play was the most popular comedy in London in Elizabethan times and it reveals a side of English life not found in Shakespeare.

In the very nature of the case it is extremely difficult to determine how far the lore and rituals of these guilds were truly initiatic. We do not know their secrets but we cannot be sure whether this is because there were no secrets or because their secrets were jealously guarded. No doubt the 'rites de passage' from the condition of layman to apprentice, from apprentice to journeyman, from journeyman to master, marked vital stages in the careers of craftworkers, in parallel to the sacraments of the Church. We have already considered the kissing of St Barbara and the ritual leap over the leather apron of the German miners and mining engineers. But whether these rites realised for those who underwent them any substantial connections between the nature of their daily work, the cosmic creation and the final redemption or release of the spirit, is not easily discoverable. The one case in which it is claimed that a genuine initiatic tradition has been preserved, masonry, is hotly contested. It is enough to say here that some of the claims made by masons are quite consistent with traditional understandings, for example that the human act of building is a repetition of the universal creation by the great architect of the universe. Given the layer upon layer of symbolism concerning the crafts in scripture, myth and folklore, it is highly probable that craftworkers made use of these symbols in their daily occupations, and developed practices to deepen their insight into them. The guild system inculcated in craftworkers a strong sense of pride in their crafts. This pride would have urged them to an understanding of their crafts at the deepest level. To think of one's craft as a symbol of the divine creation or of spiritual redemption is to see it in its most glorious aspect. Many guilds accompanied their members to their burial, covering their coffins in elaborate palls on which the instruments and symbols of their craft were embroidered.

We cannot know for certain how things stood in this regard in six-teenth century England. We do know that from the middle of this century the religious affiliations of the guilds were attacked and progressively de-stroyed by the reformers. The miracle plays were suppressed, not because they had lost their popular following but by reformist zeal, reinforced by state opposition to their alleged idolatry and superstition. We have already noted Luther's opposition to the saints' days and holy days which craftworkers regarded as important privileges, since on the days dedicated to the patron saints of each of the crafts, those craftworkers celebrated at the expense of their employers. Calvin likewise attacked the practice in his *Institutes*. In England the attempt to suppress these festivals was justified in the name of industry, just as Luther had criticised them for encouraging idleness. But it is hard to resist the feeling that the reformers had more positive objections to them, that they encouraged merry making, pranks and high spirits, and that they were superstitious. The behaviour of the apprentices in London and the various attempts to control them are matters of the greatest interest throughout this period.

In the seventeenth century reformist attacks on the religious practices of the craft guilds continued. According to the Puritans there was no authority in scripture for their worship of the saints, nor for the elaborate rituals, ornaments and vessels of the high Church. These remnants from the time before the reformation were banned or destroyed, and with them went some of the most powerful and enduring links between religion and the crafts. At this point a distinction may be drawn between southern Germany and Puritan England, since in Germany the Lutherans retained much of the medieval decoration in their churches and have preserved it to our own time. The Lutheran mining communities in southern Germany still keep paired wooden statues of miners and angels in their churches, together with screens illustrating the miners' work. They have also preserved the practice of dedicating their work to the patron saints of mining. Mathesius, an immediate disciple of Luther gave sixteen sermons to miners about mining, on the basis of St Paul's Epistle to the Philippians who were the first Christian mining community. In short, this part of Germany did not suffer from the iconoclasm which in the name of a purer religion smashed much of the finest craftwork in England. In this way we can distinguish between the Protestant and Puritan work ethics and acknowledge that the Lutherans' emphasis on traditional vocations, however much they diminished the notion, did something to preserve the connections between craft and religion. In England, however, the inspiration of the Puritan divines was Calvin, not Luther. By the time the high Church party was finally successful in 1660, the damage had been done.

There is some evidence that the Puritans themselves felt the vacuum that their iconoclasm had created, and their attempts to fill that vacuum demonstrate better than anything else the distance between the doctrines of their reformed Church and the traditional doctrine of work. Towards the end of the seventeenth century there were books published with titles such as 'Navigation Spiritualized', 'Husbandry Spiritualized', and 'The Religious Weaver'. But such works were hardly at all concerned with what we might expect from their titles, the exposition of how each of these crafts is symbolic of the divine creation or of spiritual redemption. They are concerned with saving their readers from the social vices attendant upon those forms of work: in the case of sailors, for example, from drunkenness, swearing and whoring. Nonetheless their titles promise much more, a promise which would attract readers who had some understanding of what these forms of work had meant before they came under the new dispensation.

The desacralisation of the crafts, on the ground that their religious practices were superstitious and idolatrous, was carried through by the reformers at the same time as they desacralised the land. The conversion of Christianity into a religion of the book, and the stripping away of Catholic tradition, entailed that many of the practices which sanctified the land also had to be suppressed. Once again there was no authority for them in Christ's teaching. Almost from the beginning of the reformation in England the processional beating of the bounds of the parish was severely modified. Up to this time the tradition had been that just before Ascension Day the priest and others would ceremonially walk all the way round the boundaries of his parish. The old ritual with banners and crosses and a large crowd of followers had thanked God for the gifts of the earth, had strengthened the parish against the incursion of evil spirits, and had reinforced the people's sense of the inviolability of titles and legal boundaries. But from the middle of the sixteenth century processions were banned and the beating of the bounds became instead a perambulation. The wayside crosses and the ritual drawing of crosses on the earth were done away with. Richard Corbet, who was Bishop of Oxford in the early seventeenth century, when Oxford was the centre of high Church opposition to the Puritan movement, described the effects of such reforms in a famous poem.

> Farewell, rewards and fairies,
>> Good housewives now may say,
> For now foul sluts in dairies
>> Do fare as well as they.
> And though they sweep their hearths no less
>> Than maids were wont to do,
> Yet who of late for cleanliness
>> Finds sixpence in her shoe?

At morning and at evening both
    You merry were and glad,
So little care of sleep or sloth
    These pretty ladies had;
When Tom came home from labour,
    Or Cis to milking rose,
Then merrily went their tabor,
    And nimbly went their toes.

Witness those rings and roundelays
    Of theirs, which yet remain,
Were footed in Queen Mary's days
    On many a grassy plain;
But since of late, Elizabeth,
    And, later, James came in,
They never danced on any heath
    As when the time hath been.

By which we note the Fairies
    Were of the old Profession.
Their songs were 'Ave Mary's',
    Their dances were Procession.
But now, alas, they all are dead;
    Or gone beyond the seas;
Or farther for Religion fled;
    Or else they take their ease.

Corbet was a sharp observer of his times and particularly of the effects of the Puritan revolution. He described in another poem how the old wooden crucifixes were used by Puritans as splints for horses' legs. This story of the fairies' departure reflected a common and widespread belief of the time, which often credited the West Country and then Ireland with being the last refuges of these dispossessed spirits of the land. As spirits of the land they had enabled the human inhabitants of a place to imagine the occult intelligence of the natural world around them, as though it were an extension of themselves or they an extension of it, all members finally of a single species. In another poem written nearly two centuries later, Wordsworth was to mourn the loss of the pagan gods of nature for just this reason. In this light, Milton's account of the demons which we considered in the last chapter was already an anachronism.

We have already considered the dwarfs of mining folklore, the help they gave miners, and the connections which such beliefs established between miners and the ground they worked. Similarly Corbet's poem makes very clear how the fairies played a part in domestic work and the work of the farm. These spirits of place were powerful agencies in the traditional beliefs about work. They not only rewarded the scrupulous housekeeper but celebrated the industry of the labourer and the milkmaid. Being themselves very active and alert, the fairies introduced an element of playfulness into the workaday world, the same element we have seen in the feast days, festivals, and pageants of the craftworkers' year. Their departure typifies the novelty of the Puritan attitude to work. For the Puritan, work was not an occasion for making merry, any more than it was in itself a means of spiritual development.

BLAKE AND WORDSWORTH ON WORK AND NATURE

Despite the defeat of the Puritans and the restoration of the monarchy in 1660, the damage which had been done to the traditional understandings of work and nature was not repaired. On these issues there seems to have been a compromise or a stand off between the forces of Cromwell's commonwealth and those of the restoration. It is fascinating to speculate on how far the loss of these traditional understandings contributed to the scientific and industrial revolutions which now transformed Britain and later the world. Could the chemists of the eighteenth century have done what they did if they had supposed their material to be charged with those occult powers in which their alchemical predecessors had believed? How much more difficult would it have been to establish production line processes in English factories on such a scale if manufacture had retained the aura of the sacred? Would the English midlands have become the black country if the fairies had remained? Might it not have been the removal of these restraints rather than capital accumulation, technological advances or political freedom which enabled Britain to create an entirely new human order? The old doctrines of work and nature were discarded, but was this because they were no longer in step with the new economic and political circumstances? Or were these circumstances themselves the outcome of a new spiritual order? It is just on this issue that Weber and Tawney diverge, as Marx diverged from Hegel on what governs the dialectic of history.

To these issues we cannot attend here, except to point out that the decay of the traditional doctrines of work and nature played a much greater part than is generally acknowledged. The easiest way of tracing the decay of these traditional doctrines from this point to the present time is to examine the ideas of two representative authors at the turn of the nineteenth century. One of them, William Blake, made an heroic attempt to reenact the traditional doctrine of work in his own life, but despite his personal triumph was ignored. The other, William Wordsworth, helped to establish in the English understanding an entirely new way of regarding nature, the spirit and the moral life. It is this way, devastatingly criticised by Blake, which has become a twentieth century norm and the source of many of the disputes about mining.

William Blake regarded himself as a prophet in the Old Testament manner and he remains, two centuries later, one of the most acute commentators on contemporary life. He matters here because he is the clearest exponent of the traditional doctrine of work since the Puritan revolution, both in his writing and in the way he lived. He is also the greatest poet of the metals and of metallurgy in the English language, as well as being an engraver on metal by profession and the inventor of an entirely original method of printing from metal plates. His poetry is characteristically complex and difficult; it is also a treasury of traditional doctrine or, as Blake called it, the wisdom of ages. One of his last and greatest works was called *Jerusalem,* that visionary city which he believed would be the final apotheosis of London where he lived.

Blake's *Jerusalem* is in four chapters, each of which begins with a sermon addressed to a particular group in the society of his time. The sermon at the beginning of the fourth and final chapter is addressed to the Christians:

> We are told to abstain from fleshly desires that we may lose no time from the Work of the Lord: Every moment lost is a moment that cannot be redeemed; every pleasure that intermingles with the duty of our station is a folly unredeemable, and is planted like the seed of a wild flower among our wheat: All the tortures of repentance are tortures of self-reproach on account of our leaving the Divine Harvest to the Enemy, the struggles of intanglement with incoherent roots. I know of no other Christianity and of no other Gospel than the liberty both of body and mind to exercise the Divine Arts of Imagination, Imagination, the real and eternal World of which this Vegetable Universe is but a faint shadow, and in which we shall live in our Eternal or Imaginative Bodies when these Vegetable Mortal Bodies are no more. The Apostles knew of no other Gospel. What were all their spiritual gifts? What is the Divine Spirit? is the Holy Ghost any other than an Intellectual Fountain? What is the Harvest of the Gospel and its Labours? What is that Talent which it is a curse to hide? What are the Treasures of Heaven which we are to lay up for ourselves, are they any other than Mental Studies and Performances? What are all the Gifts of the Gospel, are they not all Mental Gifts? Is God a spirit who must be worshipped in Spirit and in Truth, and are not the Gifts of the Spirit Every-thing to man? O ye Religious, discountenance every one among you who shall pretend to despise Art and Science! I call upon you in the Name of Jesus! What is the Life of Man but

Art & Science? is it Meat and Drink? is not the Body more than
Raiment? What is Mortality but the things relating to the Body
which Dies? What is Immortality but the things relating to the
Spirit which Lives Eternally? What is the Joy of Heaven but Improvement in the things of the Spirit? What are the Pains of Hell
but Ignorance, Bodily Lust, Idleness and devastation of the things
of the Spirit? Answer to yourselves and expel from among you
those who pretend to despise the labours of Art and Science, which
alone are the labours of the Gospel. Is not this plain and manifest
to the thought? Can you think at all and not pronounce heartily
That to Labour in Knowledge is to Build up Jerusalem, and to
Despise Knowledge is to Despise Jerusalem and her Builders. And
remember: He who despises and mocks a Mental Gift in another,
calling it pride and selfishness and sin, mocks Jesus the giver of
every Mental Gift, which always appear to the ignorance-loving
Hypocrite as sins; but that which is a Sin in the sight of cruel Man
is not so in the sight of our kind God. Let every Christian, as much
as in him lies, engage himself openly and publicly before all the
World in some Mental pursuit for the Building up of Jerusalem.

Elsewhere Blake wrote:

A Poet, a Painter, a Musician, an Architect, the Man or Woman
who is not one of these is no Christian.

These statements set out the major tenets of the traditional doctrine of
work in a way which perhaps only became possible at a time when this
doctrine was obsolescent. Blake can say what he says here because the
position which he expounds is already only one view among others of what
work is or should be. To an extent Blake is recapitulating teachings which
he had inherited. He mentions the duty of our station in a way which is
reminiscent of Luther, and his talk of the real and eternal world of which
this vegetable world is but a shadow is Platonic. The emphasis on mental
studies and performances in connection with liberty recalls the free act of
contemplation. In the traditional doctrine, but not in this passage of Blake,
this is contrasted with the servile act of manufacture. In the opening lines
of the sermon Blake sets out the ancient doctrine that the greatest moral
evil is dissipation, the wasting of one's time and talents on idle pleasures
and pursuits. This evil, it seems for Blake, is not usually a deliberate
turning away from one's proper work, but an incapacity to make clear to
oneself what exactly that work is. He describes this in a wonderful phrase
as 'the struggles of intanglement with incoherent roots.'

But what of Blake's claim that according to the Gospel, art and science are the real work of the spirit? Blake feels the need to argue for this

and he leads up to it carefully by referring to those passages in the New Testament which he takes to support his case, but without making it immediately clear how he intends to use them. Leaving aside the apostles for a moment, we may ask what justifies his claim that the holy ghost is the origin of the ideas which artists contemplate. We remember in this context the spirit of God which moved on the face of the waters in the first verses of Genesis and that same spirit with which God filled Bezaleel, Aholiab and the other craftworkers who made the tabernacle, its altars and its vessels. But is this the spirit which descended on the disciples at Pentecost and gave them the gift of tongues? Blake believed that it was, on the ground that this divine visitation conferred the art and science of poetry on those who received it. For Blake, Jesus too was essentially a poet.

In the case of the hidden talent, there is no reason to suppose that Jesus meant what Blake and we mean by this word. For Jesus it meant simply a certain weight of precious metal. But this word has come to mean what we mean by it because Jesus uses it as he does in the parable, even though it is impossible to tell what Jesus himself intended by the metaphor. As for the treasures in heaven, we would normally take these to be the moral virtues. But Blake distrusted the moral virtues, seeing in them little more than an occasion for accusing others of sin. He was always working to restore what he thought to be the intellectual power of Christianity, as opposed to its morality and its devotional practice. He writes later in *Jerusalem:*

> I care not whether a Man is Good or Evil; all that I care is
> Whether he is a Wise man or a Fool. Go, put off Holiness
> And put on Intellect.

Blake lived what he wrote. Overwhelmed by the delights of the intellect and imagination he neglected all other considerations. Not for him the outward prosperity and reputation of the Puritan as the sure signs of God's grace. For Blake poverty and the stigma of madness were as nothing compared to the happiness of doing his own work for its own sake. If other people failed to appreciate that work, Blake knew that the angels in heaven were delighted by it just as he was. His output was prodigious: from childhood to the songs he sang on his deathbed he lived his seventy years in an unremitting fury of creation. There was never any money. In his later years he owned a single rusty black jacket which he never wore indoors but preserved for when he had to go out. Having no servant, he often embarrassed his friends by greeting them in the street as he carried his own jug of porter back from the public house. If his poor wife Catherine told him they were penniless he would fly into a rage, so she learnt to present him with an empty plate at dinner to show him he would have to earn a commission by engraving someone else's designs. His disciple, the great painter Samuel Palmer, wrote of him that he ennobled poverty and made two little rooms off the Strand more attractive than the threshold of princes.

For Blake as for the Greeks and Jews the paradigm of creative activity was metallurgy. In his earlier work he represented the creation of the physical universe as the making of the mundane egg by means of anvils and furnaces. We may see in this some recollection of Ptah's creating the world as an egg. In his famous poem 'The Tyger' Blake wrote:

> And what shoulder and what art,
> Could twist the sinews of thy heart?
> And when thy heart began to beat
> What dread hand? and what dread feet?

> What the hammer, what the chain?
> In what furnace was thy brain?
> What the anvil, what dread grasp
> Dare its deadly terrors clasp?

But though the imagery of the smithy abounded in the early work as symbolic of universal creation, it was not much elaborated. According to some scholars, a great change came over Blake's work about half way through his career, after which the metallurgical imagery was far more potent. They suggest that Blake must actually have seen a casting mill in operation, since from this point the imagery of smelting was both more specific and more widely applied. Now we hear of the glare and roar of the fire, the clatter of hammers and blowing of bellows, the clinkers, the rattling chains, the ladles carrying molten ore, the dark gleam of the ashes still burning before the iron doors of the furnace. These are the tremendous images which Blake now used to illuminate the processes of the imagination and of the human body.

The protagonist in Blake's highly original mythology of creation is Los, whose name is probably an anagram of the Latin word for the sun, and who represents the divine imagination. He is the intellectual sun who creates the worlds of space and time and also the worlds of the imagination. Almost always he accomplishes this work by forge and furnace, and we must ask why Blake conceived of his own creative work as a poet and painter in these images. Partly it has to do with energy, which Blake often compared to the fires of hell and which he always imagined as burning. He may also have thought of his poems and designs as unbreakable, so well composed that the bonds which bound each of them in its integrity were as strong or stronger than iron or steel. Certainly they have endured. Then again, despite what has just been said, Blake was a tireless reviser of his longer works, recasting the patterns and contexts of the various passages which comprised them. Though many of the same passages appear again and again in these works, their relations to each other are changed. This process Blake seems to have thought of as a refinement, as Los throws back the earlier work into the furnace that it may emerge purified and be formed anew.

But the work of Los does not end with the creations of the poet and the painter. His metallurgical labours are also the processes of the human body.

> In Bowlahoola Los's Anvils stand and his Furnaces rage;
> Thundering the Hammers beat and the Bellows blow loud,
> Living, self moving, mourning, lamenting and howling incessantly.
> Bowlahoola thro'all its porches feels, tho' too fast founded
> Its pillars and porticoes to tremble at the force
> Of mortal or immortal arm; and softly lulling flutes,
> Accordant with the horrid labours, make sweet melody.
> The Bellows are the Animal Lungs; the Hammers the Animal Heart:
> The Furnaces the Stomach for digestion: terrible their fury.
> Thousands and thousands labour, thousands play on instruments
> Stringed or fluted to ameliorate the sorrows of slavery.
> Loud sport the dancers in the dance of death, rejoicing in carnage.
> The hard dentant Hammers are lull'd by the flutes' lula lula,
> The bellowing Furnaces blare by the long sounding clarion,
> The double drum drowns howls and groans, the shrill fife shrieks
>   and cries.
> The crooked horn mellows the hoarse raving serpent, terrible but
>   harmonious:
> Bowlahoola is the Stomach in every individual man.

This fantastic evocation of the animal organism shows how Blake conceived of the natural world. It is a continuous and infinite miracle in which the tiniest particles, the minute particulars, are organised in accordance with the will of divine powers. More than anyone else, Blake seems to have found his way to the threshold of creation where he could observe the making of thoughts and things in the realm of the invisible. In the ameliorating of the painful processes of the body by music, there is perhaps a prevision of the physiological theory that the pain occasioned by these organic processes is moderated by anaesthetic secretions of the glands.

It is always a shock to turn from the fury of Blake to the passivity of Wordsworth. The two men were contemporary for nearly sixty years, each knew the other's poems, they are both regarded as Romantics, and yet in their understandings of work and nature it would be hard to find two thinkers less alike. Blake at least was aware of the gulf which separated them. For Blake the natural world, the world of space and time, was the arena of the earthly struggles which he fought against himself and against others till he died. Through the work of his hands he engaged with that world as thoroughly as anyone could, and he believed in his work as the proper means to his salvation according to the way he interpreted the teachings of Christ and the Bible. And yet at the deepest level he abhorred the material world as a constant hindrance to his vision. In his own words, it was no more than the dirt on his feet; he looked through the eye, not with it. Wordsworth, on the other hand, regarded the world of the countryside as his teacher and moral guardian. A man of the middle classes and sustained throughout his life by sinecures and bequests, he worshipped the natural world with a fervour which Blake considered idolatrous.

For all the achievement of his early years Wordsworth seems to have known little of work, even of his own. He had doubts about whether his art would earn him a livelihood and he was uncomfortably aware that such a career might begin happily enough but often ended in disaster. He does not appear to have considered himself a poet born, with a destiny to fulfil. He was on occasion very doubtful of the value of books compared to the direct influence of nature. He lacked a sense of himself as a poet and this probably helped his readers to identify themselves with him. He was a man speaking to men, one who had divested himself of all the poetic artifice of the past in order to speak everyone's language. Wordsworth claimed nothing special for himself, no peculiar gift or talent, only that heightened sensibility which elevates anyone who has it in whatever walk of life. He presented himself as a person in no way distinguishable from his fellows by any predisposition for a particular kind of work. But if the moral life did not consist for Wordsworth in the doing of one's own work to the best of one's ability, in what did it consist? Instead he wrote of

> that best portion of a good man's life,
> His little, nameless, unremembered acts
> Of kindness and of love.

There is nothing to be said against little acts of kindness and love. But to claim that these constitute the best portion of a moral life is, in my view, staggeringly wrong. When we attempt to measure the distance between this claim and the spiritual ideal of work which Blake sets out in his sermon to the Christians, we realise that we are in two different worlds.

Wordsworth believed that merely to have lived among the works of nature in the country had called forth and strengthened his powers of imagination in boyhood and youth. In this way, according to Wordsworth, the external world sustained inward vision. That his inward vision was very powerful in his early years we cannot doubt. He described later in life how he had found it almost impossible to believe in the external reality of the world, and many times on his way to school had had to grasp hold of things to convince himself that they were not the projections of his own mind. He called the vertigo which this feeling induced in him 'the abyss of idealism'. In one of his poems he described how a visitor to a waterfall had to distract his own mind by mathematical calculations, so deeply was he affected by what he was seeing. In this visitor we may catch a glimpse of Wordsworth himself. At the first level, then, Wordsworth had visionary capacities quite as great as those of Blake, but where Blake gladly committed himself to them, even at the cost of seeming mad to those around him, Wordsworth feared them.

As a result there is little in Wordsworth's poetry which is truly visionary, nor did he investigate the source of his ideas in the way that Blake did. Instead he created for himself a dependency upon the forms of the natural world, which at once evoked and curbed his visionary powers. He remained aware that much of what he had seen in his early years he had himself envisioned, and his greatest poems detail the gradual closing of these visionary springs which had transformed, as he believed, the world of his childhood. But he was incapable of creating entirely from within himself as Blake could. His poems were typically the product of meditations in which he recollected past experiences in tranquillity. The daffodils which he once saw are brought to life in his mind as he lies upon his couch, but they are those daffodils by that shore. Blake's tiger, on the other hand, is all tigers and none of them, the veritable first tiger which God himself made. When Blake read in a copy of Wordsworth's poems Wordsworth's claim that natural objects strengthened his imagination, he wrote in the margin that natural objects always had and still did weaken, deaden and obliterate his own imagination. Wordsworth, he adds, must know that what he writes valuable is not to be found in nature.

Wordsworth's belief that natural objects strengthened the imagination is the simplest and most plausible explanation of his failure as a poet in middle age. Where Blake continued to design and write with genius to the end of his long life, Wordsworth lost his gift. He died at the age of eighty, having produced little of note during the latter half of his life. The source of Wordsworth's power as a visionary poet lay in his capacity to remember the inner events and experiences of his childhood. This was a finite stock and in any case further and further from him as he grew older. For Blake, on the other hand, the source of creation was imagination, not memory, and the ideas upon which he drew were infinite and inexhaustible. Wordsworth alone among the Romantic poets did not know his way to these waters of life in which the poetic genius is continually renewed. And so he could have no understanding of the traditional doctrine of work, in which the invisible ideas of what is to be done or made are central. Wordsworth's flight into nature and away from vision disabled him morally and in the spirit to an unusual degree. This same flight into nature is one of the great spiritual problems of our own time.

For there can be no doubt that the Wordsworthian view has triumphed. His is the moral order in which we now live, where the doing of one's work is of little or no moral significance, while kindness is all. This is the victory of niceness, for which all other moral values have been discarded and the high spiritual aspirations of previous generations forgotten. It is hard not to believe that Blake was thinking of Wordsworth when he wrote:

> He smiles with condescension, he talks of Benevolence and Virtue,
> And those who act with Benevolence and Virtue they murder time
>   on time.
> These are the destroyers of Jerusalem, these are the murderers
> Of Jesus, who deny the Faith and mock at Eternal Life,
> Who pretend to Poetry that they may destroy Imagination
> By imitation of Nature's Images drawn from Remembrance.

Such people are the destroyers of Jerusalem because Jerusalem, as Blake explained in his sermon, is the spiritual city which is built by the labours of inspired artists. We can appreciate why Blake should consider benevolent people to be murderers only when we set this new morality against the traditional doctrine of work. In Blake's view benevolence had usurped the place of art.

The shift in attitudes to which these lines of Blake point is, of all historical changes, the one most relevant to the place of mining in contemporary society. This shift is the culmination of those departures from the traditional doctrine of work which we have traced through the renaissance, the reformation and the Puritan revolution of the sixteenth and seventeenth centuries. But what distinguishes the Wordsworthian morality from those which preceded it, is that it was almost entirely secular. For Wordsworth the most important part of the moral life was not the realisation of the divine within us, as the traditional doctrine of work proposed, nor even the glorification of God as the Puritans believed. At best there was the possibility of perceiving the divine through the contemplation of natural objects. But this was idolatrous, as Blake saw, when it turned into the worship of nature as divine in itself. Nonetheless there can be no question but that the time for this new morality had come. The Wordsworthian view was already the standard view or became so very quickly, and was to be found in the other most widely read moralist in the English language in the nineteenth century, Charles Dickens. For Dickens, too, kindness was all, and work had almost no part in the moral order.

We are now in a position to see all the way around the Wordsworthian conception of nature, and to appreciate it fully. This conception of nature is so deeply a part of our thinking that it takes an act of violence to that thinking to free us from it long enough to look at it. Essentially, the Wordsworthian nature is pure spectacle: it is perceived by the eye and the ear but it is not touched. We are cut off from it as by a screen, and even the peasants who stand on the other side of the screen, though they may sing, hardly ever work. The rustic is superior to the citizen, but not because he is engaged in work which employs his intellect as well as his body, while his counterpart in the city has been transformed by the factory system into a mere hand. The rustic is superior because his passions are incorporated with the great and permanent forms of nature and because he is, like the poet, more sensitive to the simple beauties around him. He too is a spectator, and the irony is that he should have seemed so to a poet and a nation which inhabited one of the most highly cultivated landscapes in the world.

Wordsworth's views of nature and of the moral life were sentimental. They were the views of a class which had already forgotten how the wealth on which it lived had been produced, and how traditional forms of work had sustained the spirit of those who engaged in them. This is not to say that anyone who respects nature is soft in the head. Blake, too, passionately denounced cruelty to animals and the wanton destruction of the natural world. But Blake also wrote:

> The cut worm forgives the plough.

More particularly Blake did what he could to resist the new productive processes which displaced the traditional forms of work. This for Blake was where the evil lay. For Wordsworth it was the city which was evil because it was noisy, dirty and crowded, and because it was cut off from the great and permanent forms of nature. The city grated on Wordsworth's sensibility which had been most delicately attuned to the beauties of natural objects. On the one occasion in which he was tricked out of his fear by the city's beauty at dawn, his very amazement shows how settled was his conviction of its evil. Blake lived in the middle of London almost all his life, sustained entirely by the beauties of his own imagination and by his belief in the value of his work, despite his own stark vision of the city's horror.

But Blake was a man unknown. In his later years as he composed his prophetic books, he was far outside the intellectual world of his time. These were the years of deepening neglect as his first biographer called them. Meanwhile Wordsworth's star was in the ascendant. Common enough before he wrote, his view of nature was more and more widely accepted during these years. It is no coincidence that the English came to believe in the value of untouched nature at the very same time as new means of production were being introduced which ravaged the natural world to an unprecedented degree. In one important respect these events were complementary, not antithetical. As the traditional views of work and nature were lost, so the new modes of production and the Wordsworthian view of nature appeared. From then on, it looked as though the belief in untouched nature was a reaction to the devastation, but in fact, historically and psychologically, it was just another aspect of the same loss. To think of untouched nature as perfect is, from the traditional point of view, as mistaken as to think of it as an expendable resource. Nature is neither a goddess to be worshipped from a distance, nor a whore to be used up and dismissed. Nature is the wife of the human genius: together they are to produce their manifold creations to the simultaneous fulfilment of both.

Neither the love of nature nor the pastoral tradition began with Wordsworth. Poets have been idealising country life and ignoring the work of the peasant since the times of Theocritus and Virgil. But they were very different from Wordsworth. It is instructive in this connection to juxtapose how Queen Marie Antoinette played at being a shepherdess and the high moral tone of the Wordsworthian pastoral just a few years later. Before Wordsworth the pastoral tradition had existed side by side with deeper moral and spiritual codes, with the transcendent God of theology and with the traditional doctrine of work. But at this time, as Blake saw, the new morality of nature and kindness began to displace these older teachings entirely. It replaced them with a divinised, untouchable nature and an unfocussed benevolence.

EPILOGUE

THE NUCLEAR AGE

T HE DISCOVERY THAT THE nuclear metals have special properties profoundly altered the scientific conception of these metals and of matter in general. That radiation was emitted by uranium, polonium and radium showed that in these forms matter was not quite as inert as had been thought. It showed also that atoms were not the ultimate constituents of the material world. When these radiations penetrated other materials, they rendered them transparent in a way never before imagined. When Madame Curie succeeded in isolating radioactive elements in a pure form, they shone with their own light. In all these ways the discovery of the special properties in these metals broke the conceptual limits within which the physical sciences of the eighteenth and nineteenth centuries had organised themselves. The apparent solidity and opacity of the material world were suddenly removed, and the old laws and principles had to be discarded.

In a famous passage Sir James Jeans described what constituted a plank of the floor on which he stood from the point of view of the nuclear physicist. His account of that whirling mass of nuclear particles strips away the reliable, comfortable surface of things and leaves us in a profound uncertainty, as he recognised. To an extent classical atomic theory had done the same thing as far back as Democritus, but in the new physics the instability of matter was accelerated to an inordinate degree. The unseen world of matter beyond appearances had never appeared so charged and dynamic as when Einstein formulated his equation between mass and energy. The convertibility of mass into energy made the material world insubstantial in a quite new way. If this world was not the creative play of divine ideas as the ancient theologians had supposed, it was nonetheless utterly different from what our senses told us.

There are some striking parallels between this new scientific order and some of the earliest periods we consider in this book. This is less surprising than it may seem when we remember that the scientific revolution of the eighteenth and nineteenth centuries had overturned the traditional conception of the material world in much the same way as the nuclear revolution now overturned the scientific understanding of the enlightenment. The wheel which had turned half circle with the first scientific revolution completed its circuit with the second, and many of the features of the traditional view were restored. The luminous glow emitted by radioactive substances evoked awe in those who saw it. The word hyperphosphorescence was coined to describe the phenomenon. This awe has much in common with the wonder felt by the people of much earlier times when they viewed the shining of the precious metals. Even the idea that earthly substances might shine with their own light was prefigured in the Jewish belief that the stones of Aaron's breast plate glowed before a military victory. It is in the context of sacred objects that we find a parallel to the nuclear metals.

As the dangers of handling radioactive materials became better known, they acquired a still greater mystique, especially when it was discovered that they transmitted their radioactivity to everything around them. In the popular view they became not merely awful but dreadful, despite their use in medicine and radiography. The belief that certain metals are not to be touched on pain of injury or death is a very old one, most commonly found in the widespread taboos on iron. We will consider some instances of a similar idea in the consecration of the tabernacle and its furnishing. So sacred was the ark of the covenant that even the golden poles by which it was carried were sacrosanct and could be touched only by consecrated hands on pain of death. When the vessels of the temple were profaned by the feasters at Belshazzar's feast, the hand of God wrote out their doom on the plaster of the wall and that very night the kingdom of Babylon was destroyed. Those who destroyed it took great care to return the sacred vessels to Jerusalem and to rebuild the temple which housed them. These stories must have developed a special dread of these sacred things which has a counterpart at the profane and material level in the fear of the nuclear metals.

Other connections between the nuclear age and the spiritual traditions of east and west were made at the time of the first nuclear explosion. That first test in the New Mexican desert in 1945 was given the code name Trinity by the scientist in charge, Robert Oppenheimer. The reasons for that choice of name have been a matter of dispute. On one view the name referred simply to the three nuclear bombs being made at that time, the one used for the test and the two to be dropped on Japan. Another explanation is that the test site was somehow associated with an abandoned turquoise mine near Los Alamos which was called Trinity. This mine had been laid under a curse and deserted by the superstitious Indians. A third possibility is one suggested by a conversation with Oppenheimer himself at a much later date, in which he indicated that he had chosen the name Trinity because he was reading the poetry of John Donne. This remark has been taken to refer to one of Donne's Holy Sonnets:

Batter my heart, three person'd God; for, you
As yet but knocke, breathe, shine, and seeke to mend;
That I may rise, and stand, o'erthrow mee,'and bend
Your force, to breake, blowe, burn and make me new.
I, like an usurpt towne to'another due,
Labour to'admit you.

On this account Oppenheimer seems to have suggested that the test explosion was a means by which God would finally overcome the resistance of the fallen soul and restore it to himself.

Did Oppenheimer really believe this? So far as nuclear theory went, he had reason to suppose that a nuclear blasting of the human frame would reorder its constituent particles. These would persist, but the theory gave him no warrant for supposing that anything else would. But Oppenheimer was not only a student of John Donne. He was also a reader of the Hindu Bhagavad Gita, parts of which he was quoting to himself at the time of that first explosion. He compared the brilliance of the explosion to the Gita's description of God as like the radiance of a thousand suns, and he repeated to himself a line of Lord Krishna's from the same poem.

I am become death, the destroyer of worlds.

All this strengthens the possibility that the name of Trinity alluded to John Donne's poem, and that Oppenheimer conceived of the explosion in Christian as well as Hindu terms. For him, it seems, nuclear weaponry provided a kind of initiatic purging through which certain souls might pass on their way to God.

Did Oppenheimer suppose that this new bomb might lead eventually to the end of the world, as this was conceived in either the Hindu or Christian traditions? The quotation from the Bhagavad Gita

I am become death, the destroyer of worlds

suggests that he did, but this is a very fragile basis on which to ascribe such a belief to him. We often quote from or allude to works of literature, without thereby subscribing to the general view of the world which such works propound. As Oppenheimer witnessed that first explosion, which was rather bigger than most of his scientific team anticipated, this quotation may well have seemed apposite to him in some ways but not in others. But whatever Oppenheimer's opinion on this matter, many people have supposed that the development of nuclear weapons will bring about the end of the world as that event is predicted in the last book of the Bible, the book of Revelation. Almost at the end of this book, just before the descent from heaven of the new Jerusalem, there are prophecies which describe the destruction of the armies of Gog and Magog by fire sent down from God, and the casting of the devil into a burning lake. But, of course, the interpretation of the prophecies in Revelation has been a matter of dispute for many centuries, and there are massive differences between the competing views.

Much more common than the belief that these weapons have a part to play in sacred history is the fear that they will simply annihilate all living creatures and even destroy the earth itself. It is feared that this technology which the human race has developed may finally destroy the human race and all other species as well. From the spiritual point of view this concedes much more than should be conceded. It is an explicit principle of the Christian faith and of other religions that we on earth do not owe our existence to events on the material plane only, but to certain spiritual causes. This is the meaning of 'Our Father who is in heaven'. Furthermore, in this regard the relations between spirit and matter are not reciprocal. The spirit, which has no particular situation in time or space, moulds the universe into the muliform life around us. The destruction of some or all of these creatures at any one time will in no way diminish the informing spirit, any more than the destruction of a painter's work destroys the painter. Material events cannot affect these spiritual causes, and so there is no reason to fear that a nuclear holocaust will diminish in the slightest those informing powers which maintain the world. It is only because our science has lost sight of these spiritual causes that we can entertain this notion that life would be permanently annihilated by a nuclear holocaust. It is particularly sad that some churchmen used this fear of annihilation in their campaign against nuclear weapons since in doing so they adopted the limited view of the world offered by nuclear physics, and betrayed their own spiritual traditions.

Those who fear this annihilation of life by a nuclear holocaust are usually unaware of the long tradition in the west concerning the end of the world. They fear that the human race now stands on the brink of a catastrophe never before imagined. In fact, the prospect of a universal conflagration which destroys the creatures on earth is a standard anticipation of the future in both the classical and Christian traditions. There is the passage of Revelation which describes the lake of fire into which will be cast the devil and everyone else whose names are not written in the book of life. In the classical world, Plato proposed in a myth that the world was periodically destroyed by fire or flood. It was the common belief of his time that the human race was then in the iron age, the last of the four ages, so it is possible that Plato thought a general dissolution imminent. The idea of a universal conflagration at periodic intervals was taken up by the Stoics who believed that at a certain point in the historical cycle everything was converted into fire, out of which a new world order would proceed, identical in all respects to those before and after it.

From this point of view the prospect of a nuclear holocaust is merely the continuation of thoughts which have long preoccupied the minds of the west. In accordance with the increasing materialism of the western mentality, the nuclear metals and the prospect of a nuclear holocaust reinstate ancient preoccupations in more material terms. It is as though we in the west are bound to think these same thoughts, however different the religious or scientific contexts within which they are conceived. If we do not share the awe and dread which our ancestors felt for the vessels of the tabernacle, we have instead the proof that radiation can be harmful. If we no longer conceive of the end of the world in the context of a sacred history, we have nonetheless the fear of a nuclear holocaust. The preacher of Ecclesiastes claimed that there was no new thing under the sun, and the early Greek philosopher Xenophanes wrote

> If God had not created yellow honey, they
> would say that figs were far sweeter.

If our ancestors had less obvious reasons than we to fear a universal conflagration, it does not follow that they feared it less. But they conceived of it in the larger context of a sacred history or an eternal return, and in this at least they had the advantage of many of us. Aesthetically, there is a certain symmetry in the thought that nuclear metallurgy may bring about the end of this world as the metallurgy of Hephaestus created it. In this way the end cancels the beginning, but Hephaestus remains.

Possessors of a much envied happiness in learning nature's hidden mysteries, and communing in solitude with the rocks, her mighty sons... It is enough for the miner to know the hiding places of the metallic powers and to bring them forth to light; but their brilliance does not raise thoughts of covetousness in his pure heart. Untouched by this dangerous madness, he delights more in their marvellous formations, the strangeness of their origin, and the nooks in which they are hidden... His business cuts him off from the usual life of man, and prevents his sinking into dull indifference as to the deep supernatural tie which binds man to heaven. He keeps his native simplicity, and sees in all around its inherent beauty and marvel... In these obscure depths there grows the deepest faith in his heavenly Father, whose hand guides and preserves him in countless dangers... He must have been a godlike man who first taught the noble craft of mining, and traced in the rocks so striking an image of life.

Novalis

NOTES

Chapter 1. *Origins*
Pages
3.	l.20 ff. Ovid. *Metamorphoses*, Bk. I.
4.	l.22 ff. *The Homeric Hymn to Demeter.*
5-6.	l.5 ff. Pliny. *Natural History*, Bk. XXXIII.I. 1-8.
6	l.29-30. Tao Te Ching.
6-7.	l.36 ff. Thomas More. *Utopia*, Bk 2.
7.	l.24-33. Thomas More, *Utopia*, p. 88
8-10.	l.3 ff. Georgius Agricola. *De Re Metallica*, Bk. 1.
10.	l.10-12. Agricola. *De Animantibus Subterraneis (De Re Metallica,* p. 217).
	l.33-38. William Wordsworth. 'The world is too much with us'.
11-13.	l.17 ff. Ernest Ailred Worms. *Australian Aboriginal Religions.*
12-13.	l.30-1.7. Geoffrey Blainey. *Triumph of the Nomads*, pp.182-3.
14.	l.1-42. Camara Laye. *L Enfant Noir* (cited in Titus Burckhardt, *Alchemy,* pp. 14-15).
	l.44. ff. Georgius Agricola. *De Re Metallica*, Bk 1.
15.	l.16 ff. Diodorus Siculus III, 12-14, v.35-8.
16-17.	l.31-1.9. Alfred Zimmern. *The Greek Commonwealth.* pp. 401-402.
17.	l.24 ff. William H. Prescott. *The Conquest of Peru*, Bk.I, Ch.II, pp. 31 ff.
18 ff.	Cedric E. Gregory. *A Concise History of Mining*, pp. 196-8. Wolfgang Paul. *Mining Lore*, pp. 198-200. *Cambridge Economic History of Europe.* pp.450, 456.
19.	l.13-35. Job 28: 1-11.

Chapter 2. *The Classical Tradition*
21.	l.1-17. Homer. *Odyssey*, Bk. VII.
22.	l.22-27. Homer. *Odyssey* Bk. XI, *Iliad,* Bk. XV.
23.	l.16-18. Homer. *Iliad,* Bk.VI, XXII.
	l.22-24. Homer. *Iliad,* Bk. XXII, lines 317-20.
	l.29 ff. Homer. *Odyssey*, Bk. VII.
24. ff.	l.21 ff. Homer. *Iliad,* Bk. XVIII.
25.	l.1-2. Homer. *Iliad,* Bk. XVIII, lines 401-2.
26.	l.11-16. Homer. *Iliad,* Bk. XVIII, lines 373-379.
	l.22 ff. Proclus. *Apology for the Fables of Homer* (Thomas Taylor).
27.	l.14 ff.Plato. *Timaeus.*
28.	l.37-42. Homer. *Iliad,* Bk. XVIII, l.417-422.
29.	l.25 ff. Xenophanes. *Frag. 25* (Freeman). Heracleitus. *Frags.41, 64* (Freeman). Proclus. *Apology for the Fables of Homer,* XI, XIV (Thomas Taylor). Simone Weil. *The Need for Roots*, p.282.
30.	l.8. William Blake. *Preface to Jerusalem*, lines 38-40.

	1.9. Gal.2:20.
31.	1.9. ff. Homer. *Iliad*, Bk. I. Bk. XVIII.
	1.41-43. Homer. *Iliad*, Bk. XVIII, 1.483-6.
32.	1.11-14. Empedocles (Freeman).
32.	1.28-39. Homer. *Iliad*, Bk. XVIII, lines 590-605.
33.	1.34-35. Aeschylus. *Prometheus Bound*, lines 450-452.
34.	1.8-10. Aeschylus. *Prometheus Bound*, lines 498-500.
35.	1.11 ff. Ovid. *Metamorphoses*, Bk. VI.
36.	1.7-8. Ovid. *Metamorphoses*, Bk. VI.
37.	1.4 ff. Proclus. *Platonic Theology*, Vol. 1, Bks. I-III.
38.	1.5 ff. Plato. *Republic*, Bk. X.
	1.29 ff. Plato. *Republic*, Bk. VII.
40.	1.2-6. Plato. *Republic*, Bk. II.
40-41.	1.24-5. Plato. *Phaedo*, 110-111.
41.	1.21 ff. Plato. *Timaeus*, 33b.

Chapter 3. *The Biblical Tradition*

42.	1.1-15. Ex. 25:31-40.
	1.32-33. Deut. 8:9.
43.	1.6 ff. eg. Clement of Alexandria. Stromata, Bk. I, Chap.xxi. (Anti-Nicene Fathers).
	1.14 ff. Walter Beltz. *God and the Gods*.
44.	1.7 ff. Herodotus. *Histories*, Bk II, Chaps. 43, 50ff.
	1.15 ff. *The Book of the Dead*. E.A.Wallace Budge. *The Gods of the Egyptians*.
45-46.	Budge. *The Gods of the Egyptians*. Cyril Aldred. *Jewels of the Pharoahs*, p. 19.
46.	1.21 ff. Gen.1.
	1. 39 ff. Gen.2.
47.	1. 3-8. Gen.2.
	1. 11-23. Gen.6,7,8.
48.	1.2 ff. Gen.4.
	1.10 ff. *Chumash with Targum Onkelos, Haphtaroth*, and *Rashi's Comm.*, Vol. I, Bereshith.
	1.15 ff. Gen.4.
	1.22 ff. Giordano Bruno. *De Umbris Idearum*.
	1.37-42 ff. Gen.4:20-21.
49.	1.8 ff. Babylonian *Talmud*. *Chumash with Targum Onkelos, Haphtaroth* and *Rashi's Commentary*.
50.	1.36 ff. Robert Graves and Raphael Patai. *Hebrew Myths: The Book of Genesis*. Walter Beltz. *God and the Gods: Myths of the Bible*.
51.	1.13-15. Deut. 4.
	1.41 ff. Ex. 19.
52.	1.5 ff. René Guénon. The Mountain and the Cave (Studies in Comparative Religion).
	1.10-11. Ex. 3, 4.
	1.11-12. Gen. 8.

l.35 ff. Ex. 19, 20.

53.   l.6 ff. Ex. 20.

l.19 Ex. 20:25.

l.21. Ex. 25.

l.24 ff. *Chumash with Targum Onkelos, Haphtaroth* and *Rashi's Comm.*, Vol. II, Shemoth. *The Torah:* A Modern Commentary.

54.   l.8 ff. Ex. 25.

l.42-43. Ex. 25:20.

55.   l.1 ff. Clement. *Stromata*, Bk. V, Ch. VI (Anti-Nicene Fathers). Philo Judaeus. *Who is the Heir,* 217. *Talmud.* Numbers Rabbah 15:9. Ex. 25.

l.32 ff. *Chumash with Targum Onkelos, Haphtaroth* and *Rashi s Commentary.*

56.   l.12 ff. Josephus. *The Jewish Wars*, Bk. VII, Ch.v, part v.

l.19 ff. Josephus. *The Antiquities of the Jews*, Bk. III, Ch.vi, part 7.

l.23 ff. Philo Judaeus. *Who is the Heir*, XLV, 221 ff. Clement of Alexandria. *Stromata*, Bk.V, Ch.vi.

57.   l.11 ff. Josephus. *The Antiquities of the Jews*, Bk. III, Ch.vii, part 7.

l.6 ff. Ex. 25, 26, 30

l.22 ff. Ex. 28.

l.27-29. Babylonian *Talmud.* Tractate Gitten 68B.

l.32-37. Josephus. *The Antiquities of the Jews.* Bk. III, Ch.viii, part 9.

l. 39-40. Ex. 34:29-35.

58.   l.3 ff. Josephus. *The Antiquities of the Jews*, Bk. III, Ch.vii, part7. Philo Judaeus. *Moses* II, 122.

l.10-11. *The Life of Flavius Josephus* (Josephus: Complete Works).

l.14 ff. Ex. 28.

l.26-27. Josephus. *The Antiquities of the Jews*, Bk. III, Ch. viii, pt. 7.

l.32-43. Ex. 31:1-6.

59.   l.1 ff. Ex. 35, 6

59.   l.30 ff. Ex. 40.

l.37 ff. 1 Kings 6 ff.

l.42 ff. 1 Chron. 28, 29.

60.   l.4 ff. 1 Kings 8.

l.13. 1 Kings 8.

l.20-22. 1 Kings 5, 2 Chron. 2.

l.30 ff. 1 Kings 7.

l.35-36. 2 Chron. 9.

l.40 ff. 1 Kings 6.

61.   l.1 ff. 1 Kings 6, 7, 8.

l.12 ff. Dan. 1.

l.21 ff. Dan. 5.

l.32 ff. Dan. 5.

62.   l.7 ff. Dan. 5.

l.27 ff. Dan.2.

62-63   l.41-10. Dan. 2:31-35.

63.   l.12 ff. Dan.2.

l.29 ff. Clement of Alexandria. Stromata (Anti-Nicene Fathers).

64.   l.26-29. eg. Luke 20.
65.   l.17 ff. Hesiod. *Works and Days*, lines 109-175.
        l.33-34. Walter Beltz. *God and the Gods*, p. 209.
        l.39 ff. Plato. *Republic*, Bk.III, 415a-d.
66.   l.2 ff. *Sources of Indian Tradition*, Vol. 1, p.218-219.
        l.36 ff. Dan. 3.
67.   l.12-13. Prov. 17:3.
        l.28 ff. Dan. 3:25, 28.
        l.31 ff. *The Jewish Mystics*, ed. by Louis Jacobs.
        l.33-35. Luke 24.
        l.38-39. Dan. 8.
68.   l.1-5. Dan. 10:5-6.
        l.12-13. Matt. 27.
        l.31 ff. Num. 1, 2 ff.
69.   l.8-10. Matt. 27.
        l.24-37. Heb. 9:19-28.
70.   l.1 ff. Heb. 9.
71.   l.1-4. Henry Chadwick. *The Early Church*, Chap. 3.
        l.8-9. eg. Luke 14, Matt. 19.
        l.35. Matt. 4.
72.   l.32. Matt. 2.
        l.34 ff. Origen. *Contra Celsus* I ix (Anti-Nicene Fathers).
73.   l.7 ff. Rev. 1, 2, 3.
73-74. l.22-11. Rev. 21:9-23.
74.   l.34 ff. René Guénon. *The Reign of Quantity*, pp. 170-176.

Chapter 4. *The Medieval Synthesis*

76.   l.1-9. Abbot Suger. *De Rebus in Administratione sua Gestis*, p.65.
        l.26 ff. H.R. Ellis Davidson. *Gods and Myths of Northern Europe*,
        pp. 42-43, 80-82.
77.   l.10-14. Ex.32.
78.   l.32-34. 2 Macc.15:11-16.
        l.18 ff. *The Oxford Dictionary of the Christian Church*, p. 1227.
        l.34-35. Rom. 12, Eph. 2.
79.   l.38 ff. Jacopus de Varagine. *The Golden Legend*, pp.198-205.
80.   l.34 ff. *Oxford Dictionary of Saints*, p.31.
        l. 39 ff. Cedric E. Gregory. *A Concise History of Mining*, pp.203 ff.
81-82. l.41 ff. Wolfgang Paul. *Mining Lore*, p.457.
82.   l.15 ff. Paul. p.465-6.
        l.27 ff. Paul. p.466.
82-83. l.40 ff. *Butler's Lives of the Saints.*
84.   l.8-11. Zohar: *The Book of Splendour: Basic readings from the*
        Kabbalah. ed. by Gershom Sholem.
85.   l.27-34. Abbot Suger. *De Rebus in Administratione sua Gestis*, p.65.
86.   l.10 ff. Suger. *De Rebus in Administratione sua Gestis, Libellus*
        Alter De Consecratione Ecclesiae Sancti Dionysii, and *Ordinatio*

A.D MCXL vel MCXLI Confirmata.

87. 1.9 ff. Suger. *De Administratione*, pp.65-7.
88. 1.3-6. Suger. *De Administratione*, p.67.
  1.11-17. Suger. *De Consecratione*, p.107.
  1.18 ff. George Henderson. *Gothic*. Otto von Simson. *The Gothic Cathedral*. Irwin Panofsky. 'Abbot Suger of Saint Denis', in *Meaning in the Visual Arts*, p.109.
  1.34 ff. Panofsky. 'Abbot Suger of Saint Denis', pp.120-124. von Simson. *The Gothic Cathedral*. pp.34 ff.
89. Henderson. *Gothic*. von Simson. pp. 39ff., pp.55 ff. Panofsky. 'Abbot Suger of Saint Denis', pp.120-124, 132-133.
90. 1.9 ff. Panofsky. 'Abbot Suger of Saint Denis', p.126. von Simson. pp.103ff.
91. 1.1-14. Panofsky. 'Abbot Suger of Saint Denis', pp. 127-128.
92. 1.1-4. Panofsky. 'Abbot Suger of Saint Denis', p.131.
  1.9-14. Plato. *Republic*, Bk.VII, 528e-530e.
  1.28-38. Panofsky. 'Abbot Suger of Saint Denis', p.131.
93. 1.15. Panofsky. 'Abbot Suger of Saint Denis', p.134.
93-94. 1.31-6. Dionysius the Areopagite, *The Celestial Hierarchy*.
94. 1.8-10. Suger. *De Administratione*, p.55.
  1.20-22. Plato. *Phaedo*, 110 ff.
95. 1.27-33. Plato. *Timaeus*, 31-35.
96. 1.19-20. John 1:1-18.
  1.22 ff. Dionysius the Areopagite. *The Celestial Hierarchy*.
97. 1.40. ff. Panofsky. 'Abbot Suger of Saint Denis', p.129.
98 ff. Richard Kieckhefer. *Magic in the Middle Ages*, pp. 133, 140 ff. M.A.Atwood. *Hermetic Philosophy and Alchemy*. Titus Burckhardt. *Alchemy*. E.J.Holmyard. *Alchemy*. E.A.Wallace Budge. *Amulets and Superstitions*. Isidore Kozminsky. *The Magic and Science of Jewels and Stones*,vol.1. George Frederick Kunz. *The Curious Lore of Precious Stones*. Herbert Silberer. *Hidden Symbolism of Alchemy and the Occult Arts*. Mellie Uyldert. *Metal Magic*.
98. 1.18-20. Geoffrey Chaucer. *The Canon's Yeoman's Tale*.
101. 1.12-20. eg. Meister Eckhart.
  1.19-20. Matt. 5:3.
103. 1.1-3. David Goldstein. *Jewish Mythology*.
103-105. 1.31 ff. M.A.Atwood. *Hermetic Philosophy and Alchemy*, pp.41ff.

Chapter 5. *The Symbol of the Mine*

106. 1.1-12. Vanoccio Biringuccio. *The Pirotechnia*, pp.17-18.
107. 1.3 ff. Wolfgang Paul. *Mining Lore*, pp.164, 167-8.
108-109. 1.25 ff. Beowulf, 2127, 2128-3182.
110. 1. 24-38. J.R.R.Tolkien. *The Hobbit*. pp.197-8.
110-111. 1.43-4. Titus Burckhardt. *Alchemy*. p. 131.
111. 1.4-7. Tertullian (Anti-Nicene Fathers).
  1.11-43. *The Nibelungenlied*, chaps. 1-18.

l.25-36.*The Nibelungenlied*, p.121.

112.   l.11-19. *Larousse Encyclopedia of Mythology*, p.279. H.R.Ellis Davidson. *Gods and Myths of Northern Europe*, pp.42-4.

l.23-31. Herodotus. *Histories*. Bk 2, ch. 51; Bk 3, chs. 37-8.

113.   l.4 ff. *Larousse Encylopedia of Mythology*, p.279. Wolfgang Paul. *Mining Lore*.p.163.

l.22-25. Wolfgang Paul. p.819.

113-114. l.30-7. Georgius Agricola. *De Re Metallica*, Bk.VI, p.217.

114-115. l.12-10. Wolfgang Paul. *Mining Lore*, pp.471-4.

115-116. l.40-9. Milton. *'Il Penseroso*, lines 85-96.

116.   l.27-35. Plato. *Timaeus*.

l.35 ff. Hermes Trismegistus. *A Treatise on Initiations, or Asclepius*, pt.1 *(The Virgin of the World)*.

117.   l.23 ff. Origen. *Contra Celsus* XXIII, xxii (Anti-Nicene Fathers). E.J.Holmyard. *Alchemy*, p.21.

118.   l.1 ff. Brothers Grimm.

l.23. Apuleius. *The Golden Ass.*

Chapter 6. The Desacralisation of Work

121.   l.1-11. Hermes Trismegistus. *Asclepius* I *(The Virgin of the World)*.

122.   l.13 ff. *Bhagavad Gita*, Bks 3, 4. Plato. *Republic*, Bk.2.

123.   l.10 ff. Plato. *Republic*, Bk.2, 368 ff.

l.21 ff. *Bhagavad Gita*, Bk.3, 1 ff.

123-124. l.41-3. *Bhagavad Gita*, Bk 4, v.21.

126.   l.7-9. Mircea Eliade. *The Forge and the Crucible*.

l.15. William Blake. *The Marriage of Heaven and Hell*, pl. 10, l. 8.

127.   l.7-9. Ananda Coomaraswamy. *Christian and Oriental Philosophy of Art*, p.24.

l.20 ff. R.H.Tawney. *Religion and the Rise of Capitalism*, p.97.

128.   l.1-18. Dante. *Inferno*, Canto 11, lines 94 ff.

129.   l.34-36. Max Weber. *The Protestant Ethic and the Spirit of Capitalism*.

129-130. l.41 ff. Tawney. Weber.

130.   l.15-16. Thomas More. *Utopia*, Bk.1, p.24.

l.29-32. Benvenuto Cellini. *The Life Of Benvenuto Cellini*.

130-131. l.39 ff. Tawney. Weber.

132.   l.19 ff. Weber, chap.3.

133.   l.22 ff. E.Lipson. *Economic History of England*. *Cambridge Economic History of Europe*. S.Lilley. *Men, Machines and History*.

134.   l.6-15. Thomas Dekker. *The Shoemakers' Holiday*.

135.   l.1 ff. Tawney. pp.101-2, 107.

l.27-30. Wolfgang Paul. *Mining Lore*. pp.140-141.

l.31-33. Paul. 124-126.

136.   l.3-6. Tawney. pp.242-3.

136-137. l.35-26. Richard Corbet. *The Fairies' Farewell*.

139-140. l.22-19. William Blake. *Jerusalem*, pl.77, lines 1-50.

140.	l.23-24. Blake. *Laocoon*, k.776.
141.	l.13 ff. Matt. 25:14-30.
	l.25-27. Blake. *Jerusalem*, pl.91, lines 54-56.
141-142.	l.29 ff. Alexander Gilchrist. *Life of William Blake*, p.356.
142.	l.5-7. Blake. *Milton*.
	l.10-18. Blake. 'The Tyger' , *Songs of Experience*.
	l.21 ff. S.Foster Damon. *A Blake Dictionary*.
143.	l.12-30. Blake. *Milton*, pl.24, lines 51-67.
144.	l.3 ff. Blake. 'Annotations to "Poems" by William Wordsworth'.
	l.33-35. William Wordsworth. 'Tintern Abbey', line 34.
144-145.	Thomas de Quincey. *Recollections of the Lakes and the Lake Poets*.
145.	l.25-27. Wordsworth. 'Daffodils'.
	l.29-33. Blake. 'Annotations to "Poems" by William Wordsworth'.
146.	l.16-22. Blake. *Milton*, pl.41, lines 19-24.
147.	l.28. Blake. *The Marriage of Heaven and Hell*, pl. 7, line 6.
	l.36-38. Wordsworth. 'Composed upon Westminster Bridge, Sept. 3, 1802'.

Epilogue

149.	l.14 ff. Sir James Jeans. *The Edinburgh Lecture*.
150.	l.19-22. Josephus. *The Antiquities of the Jews*, Bk. III, ch. III, sect. 9.
	l.29 ff. Lansing Lamont. *Day of Trinity*, p. 70.
151.	l.1-6. John Donne. *Holy Sonnets* xiv.
	l.11 ff. Robert Jungk. *Brighter Than a Thousand Suns*, p.183.
151-152.	l.42 ff. Rev. 20.
152.	l.35-37. Rev. 20.
	l.38-39. Plato. *Timaeus* 22.
153.	l.14-15. Eccl. 1:9-10.
	l.18-19. Xenophanes Fr. 38. (Freeman).

BOOK REFERENCES

Chapter 1

Agricola, Georgius. *De Re Metallica.* Translated by Herbert Clark Hoover and Lou Henry Hoover. Dover Publications, New York, 1950.
Blainey, Geoffrey. *Triumph of the Nomads.* Sun Books, Melbourne, 1976.
Burckhardt, Titus. *Alchemy.* Element Books, London, 1986.
Cambridge Economic History of Europe, Vol.II. Cambridge University Press, Cambridge, 1952.
Diodorus Siculus III. Translated by C.H.Oldfather. Loeb, Harvard, 1935.
Gregory, Cedric. *A Concise History of Mining.* Pergamon Press, New York, 1980.
Hesiod, the Homeric Hymns and the Homerica. Translated by H.G.Evelyn White. Loeb, Harvard, 1914.
More, Thomas. *Utopia.* Translated by Robert M.Adams, W.W.Norton &Co., New York, 1992.
Ovid. *Metamorphoses,* Vols.I&II. Translated by Frank Justus Miller. Loeb, Harvard, 1977.
Paul, Wolfgang. *Mining Lore.* Morris Printing Co., Portland, Or., 1970.
Pliny. *Natural History.* Loeb, Harvard, 1952.
Prescott, William H. *The Conquest of Peru.* J.M.Dent and Sons, London, 1908.
Wordsworth, William. *Wordsworth's Poems.* Dent, London, 1955.
Worms, Ernest Ailred. *Australian Aboriginal Religions.* Nelen Yubu Missiological Unit, 1986.
Zimmern, Alfred. *The Greek Commonwealth.* Oxford University Press, London, 1961.

Chapter 2

Aeschylus. Vol. I. Translated by Herbert Weir Smith. Loeb, Harvard, 1922.
Bible: The New King James Version.
Blake, William. *The Complete Poems.* Alicia Ostriker (ed.). Penguin Harmondsworth, England, 1977.
Freeman, Kathleen (ed.). *Ancilla to the Presocratic Philosophers.* Basil Blackwell, Oxford, 1971.
Homer. *The Iliad, The Odyssey.* Translated by A.T.Murray. Loeb, Harvard, 1919.
Ovid. *Metamorphoses,* Vols.I&II. Translated by Frank Justus Miller. Loeb, Harvard, 1977.
Plato. *Collected Dialogues.* Edited by Edith Hamilton and Huntington Cairns. Princeton University Press, 1963.
Proclus. *Platonic Theology.* Translated by Thomas Taylor, 1816. Great Works of Philosophy Reprint, Texas, 1985.
Studies in Comparative Religion. Spring, 1971, Vol. 5, no. 2.
Taylor, Thomas. *The Works of Plato.* 1804 edition, 5 vols., published by Thomas

Taylor. AMS reprint, 1979.

Weil, Simone. *The Need for Roots.* Routledge and Kegan Paul, London, 1987.

Chapter 3

Aldred, Cyril. *Jewels of the Pharoahs.* Thames and Hudson, London, 1971.

Anti-Nicene Fathers. Edinburgh edition. Reprint by William B. Eerdman s Publishing Company, Grand Rapids, Michigan, 1981.

Babylonian Talmud. Translated and edited by I.Epstein. Torah La-Am edition, Jerusalem, 1961.

Beltz, Walter. *God and the Gods: Myths of the Bible.* Penguin, Harmondsworth, England, 1983.

Bible: The New King James Version.

Bruno, Giordano. *De Umbris Idearum.* Paris, 1582.

Budge, E.A.Wallace (translator). *The Book of the Dead.* 2nd ed. Routledge and Kegan Paul, London, 1923.

Budge, E.A.Wallace. *The Gods of the Egyptians,* vols. 1 &2. Dover Publications, New York, 1969.

Chadwick, Henry. *The Early Church.* Penguin, Harmondsworth, England, 1967.

Chumash with Targum Onkelos, Haphtaroth, and Rashi's Commentary, Vol. I, *Haphtaroth,* and Vol. II, *Shemoth.* Translated and annotated by Rabbi A.M.Silbermann and Rev. M.Rosenbaum. Bereshith, Jerusalem, 1961.

Graves, Robert, and Patai, Raphael. *Hebrew Myths: The Book of Genesis.* Arena, 1989.

Guénon, René. *The Reign of Quantity.* Penguin, Baltimore, Maryland, 1972.

Herodotus. *The Histories.* Translated by A.D.Godley. Loeb, Harvard, 1975.

Hesiod, the Homeric Hymns and the Homerica. Translated by H.G.Evelyn White. Loeb, Harvard, 1914.

Jacobs, Louis (ed.). *The Jewish Mystics.* Keter Publishing House, Jerusalem, 1976.

Josephus. *Complete Works.* Translated by William Whiston. Kregel Publications, Grand Rapids, Michigan, 1960.

Philo Judaeus. *De Vita Mosis.* Translated by F.H.Colson. Loeb, Harvard, 1935.

Philo Judaeus. *Who is the Heir?* Loeb, Harvard, 1932.

Plant, W. Gunther. (ed.).*The Torah: A Modern Commentary.* Union of American Hebrew Congregations, New York, 1967.

Plato. *The Collected Dialogues.* Edited by Edith Hamilton and Huntington Cairns. Princeton University Press, Princeton, 1963.

Sources of Indian Tradition, Vol I. Columbia University Press, New York, 1958.

Chapter 4

Atwood, M.A. *Hermetic Philosophy and Alchemy.* The Julian Press, New York, 1960.

Bible: The New King James Version.

Budge, E.A.Wallace. *Amulets and Superstitions.* Dover, New York, 1978.

Burckhardt, Titus. *Alchemy.* Element Books, London, 1986.

Butler's Lives of the Saints, concise edition. Michael Walsh (ed.).Dove Communications, Melbourne.

Chaucer, Geoffrey. *The Canterbury Tales.* J.M.Dent and Sons, London, 1958.

Dionysius the Areopagite: *Works.* Translated by J. Parker. 1897.

Eckhart, Meister. Translated by Raymond B.Blakney. Harper and Row, New York, 1941.

Ellis Davidson, H.R. *Gods and Myths of Northern Europe.* Penguin, Harmondsworth, England, 1964.

Goldstein, David. *Jewish Mythology.* Hamlyn, London, 1987.

Gregory, Cedric E. *A Concise History of Mining.* Pergamon Press, New York, 1980.

Henderson, George. *Gothic.* Penguin, Harmondsworth, England, 1967.

Holmyard, E.J. *Alchemy.* Penguin, Harmondsworth, England, 1957.

Kieckhefer, Richard. *Magic in the Middle Ages.* Cambridge University Press, Cambridge, 1989.

Kozminsky, Isidore. *The Magic and Science of Jewels and Stones.* Vol. I. Cassandra Press, San Rafael, California, 1988.

Kunz, George Frederick. *The Curious Lore of Precious Stones.* Bell Publishing Co. New York, 1989.

Oxford Dictionary of Saints. David Hugh Farmer (ed.). Oxford University Press, London.

Oxford Dictionary of the Christian Church. F.L.Cross and A.E.Livingstone (eds.). Oxford University Press, London, 1974.

Panofsky, Irwin. *Meaning in the Visual Arts.* Doubleday, New York, 1955.

Paul, Wolfgang. *Mining Lore.* Morris Printing Company, Portland, Or., 1970.

Plato. *Collected Dialogues.* Edited by Edith Hamilton and Huntington Cairns. Princeton University Press, Princeton, 1963.

Silberer, Herbert. *Hidden Symbolism of Alchemy and the Occult Arts.* Dover, New York, 1971.

von Simson, Otto. *The Gothic Cathedral.* Princeton University Press, Princeton, 1962.

Suger, Abbot. *On The Abbey Church of Saint Denis and its Art Treasures.* 2nd edn. Edited and translated by Irwin Panofsky and Gerda Panofsky-Soergel. Princeton University Press, Princeton, 1979.

Tolkien, J.R.R. *The Hobbit.* George Allen and Unwin. London, 1966.

Uyldert, Mellie. *Metal Magic.* Turnstone Press, Northamptonshire, 1980.

de Varagine, Jacopus. *The Golden Legend.* Translated by William Caxton. J.M.Dent and Sons, London, 1900. Reprinted by AMS Press, New York, 1973.

Zohar, The Book of Splendour: Basic Readings from the Kabbalah. Gershom Scholem (ed.). Schocken Books, New York, 1963.

Chapter 5

Agricola, Georgius. *De Re Metallica.* Translated by Herbert Clark Hoover and Lou Henry Hoover. Dover Publications, New York, 1950.

Anti-Nicene Fathers. Edinburgh edn. Reprint by William B.Eerdman's Publishing Company, Grand Rapids, Michigan, 1981.

Apuleius. *The Golden Ass.* Translated by Robert Graves. Penguin, Harmondsworth, England, 1950.

Beowulf. Translated by Michael Alexander. Penguin, Harmondsworth, England, 1973.

Biringuccio, Vanoccio. *The Pirotechnia.* Translated by Cyril Stanley Smith and Martha Teach Gnudi. Dover Publications, New York, 1990.

Burckhardt, Titus. *Alchemy.* Element Books, London, 1986.

Davidson, H.R.Ellis. *Gods and Myths of Northern Europe.* Penguin, Harmondsworth, England. 1964.

Grimm's Fairy Tales. Pantheon Books, New York, 1972.

Hermes Trismegistus. *The Virgin of the World.* Translated by A Kingsford and E.Maitland. Wizard's Book Shelf reprint of 1885 edn., 1977.

Holmyard, E.J. *Alchemy.* Penguin, Harmondsworth, England, 1957.

Larousse Encyclopedia of Mythology, Hamlyn, London, 1968.

Milton, John. *Paradise Lost and Selected Poetry and Prose.* Holt, Rhinehart and Winston, New York, 1951.

Nibelungenlied. Translated by A.T.Hatto. Penguin, Harmondsworth, England, 1969.

Paul, Wolfgang. *Mining Lore.* Morris Printing Co., Portland, Or. 1970.

Plato. *Collected Dialogues.* Edited by Edith Hamilton and Huntington Cairns. Princeton University Press, Princeton, 1963.

Tolkien, J.R.R. *The Hobbit.* George Allen and Unwin, London, 1966.

Chapter 6

Bhagavad Gita. Translation & commentary by Swami Chinmayananda. Published by Sri Ram Batra, Central Chinmaya Mission Trust, Bombay.

Blake, William. *The Complete Poems.* Edited by Alicia Ostriker. Penguin, Harmondsworth, England, 1977.

Blake, William. *William Blake's Writings,* Vols.I&II. G.E.Bentley (ed.). Clarendon Press, Oxford, 1978.

Cambridge Economic History of Europe. Cambridge University Press, Cambridge, 1952.

Cellini, Benvenuto. *The Life of Benvenuto Cellini, Written by Himself.* Phaidon, London, 1949.

Coomaraswamy, Ananda K. *Christian and Oriental Philosophy of Art.* Dover, New York, 1956.

Damon, S. Foster. *A Blake Dictionary.* Thames and Hudson, London, 1973.

Dante. *The Inferno.* Translated by John Ciardi and Archibald MacAllister. New American Library, New York, 1954.

Dekker, Thomas. *The Shoemakers' Holiday.* J.M.Dent & Sons, London, 1926.

Eliade, Mircea. *The Forge and the Crucible.* University of Chicago Press, Chicago, 1978.

Gilchrist, Alexander. *Life of William Blake.* Macmillan, 1880. Reprint by Rowman & Littlefield, New Jersey, 1973.

Hermes Trismegistus. *The Virgin of the World.* Translated by A.Kingsford and E.Maitland. Wizard's Bookshelf reprint of 1885 edn., 1977.

Thomas, Keith. *Religion and the Decline of Magic.* Penguin, Harmondsworth, England, 1971.

Lilley, S. *Men, Machines and History.* Lawrence and Wishart, London, 1965.

Lipson, E. *Economic History of England.* A.& C.Black, London, 1960-64.

More, Thomas. *Utopia.* Translated by Robert M. Adams. W.W.Norton &Co.,

New York, 1992.
New Oxford Book of English Verse. Clarendon Press, Oxford, 1972.
Oxford Dictionary of the Christian Church. Edited by E.A.Livingstone and F.L.Cross. Oxford University Press, Oxford, 1974.
Paul,Wolfgang. *Mining Lore.* Morris Printing Co., Portland, Or., 1970.
Plato. *Collected Dialogues.* Edited by Edith Hamilton and Huntington Cairns. Princeton University Press, Princeton, 1963.
de Quincey, Thomas. *Recollections of the Lakes and the Lake Poets.* Penguin, Harmondsworth, England, 1970.
Tawney, R.H. *Religion and the Rise of Capitalism.* Penguin, England, 1938.
Weber, Max. *The Protestant Ethic and the Spirit of Capitalism.* George Allen and Unwin, London, 1930.

Epilogue

Bible: The New King James Version.
Donne, John. *Selected Poems.* Penguin, Harmondsworth, England, 1950.
Freeman, Kathleen (ed.). *Ancilla to the Presocratic Philosophers.* Basil Blackwell, Oxford, 1971.
Jungk, Robert. *Brighter than a Thousand Suns.* Penguin, Harmondsworth, England, 1960.
Lamont, Lansing. *Day of Trinity.* Hutchinson, London, 1966.
Plato. *Collected Dialogues.* Edited by Edith Hamilton and Huntington Cairns. Princeton University Press, Princeton, 1963.

INDEX

Aaron 68, 69, 77
Aaron's breast plate 57-8, 150
Abel 48-50
Aborigines, Australian 11-13, 39
Achilles 23, 24
Adam & Eve 46-9, 99ff, 123
Aeschylus 33-4
Agricola, Georgius 8-10, 14, 113-14
Albert the Great 98
Andvari 112
Annaberg, Austria 82
Anne, St 82
Aphrodite 45
Apollo 36, 108
Apuleius 118
Arachne 35-6
Aristotle 128
Ark of the Covenant 54, 150
Athene 35-6, 84
Athenian mines 16-7, 81
Atomic test, 'Trinity' 150ff
Augustine, St 96
Augustus Caesar 3
Australia 11-13, 18
Avicenna 99
Barbara, St 79ff, 107
Belshazzar 61 ff, 150
Beowulf 108-9.
Bernard of Clairvaux 88ff
Bezaleel & Aholiab 55, 58, 60, 83, 141
Bhagavad Gita 122ff, 151
Biringuccio, Vanoccio 106
Black Death 130
Blake, Catherine 141-2
Blake, William 30, 126, 139ff
Bruno, Giordano 48
Budge, E.A. Wallace 44ff
Cain 48-50
Calvin, John 132ff
Cambyses 112
Cellini, Benvenuto 130
Charis 24ff.

Chaucer, Geoffrey 98
Christ 63, 64, 67ff, 76, 98, 100, 102, 141
Christopher, St 80
Cistercians 88ff.
Copper Mountain 114-5
Corbet, Richard 136-7
Curie, Marie 149
Daniel 61ff, 103
Dante's *Inferno* 128
David 59
De Re Metallica 8-10, 14, 113-4
Dekker, Thomas 134
Delphic oracle 108
Demeter 4
Dickens, Charles 147
Diodorus Siculus 15ff, 19.
Dionysius the Areopagite 90ff.
Donne, John 150-1
Dunstan, St 82-3
Ecclesiastes 153
Egyptian *Book of the Dead* 44ff.
Egyptian mines 15ff
Einstein, Albert 149
Eureka Stockade 18
Fafnir 111
Flammel, Nicholas 103-5
Florence 130
Gaia 11
Garden of Eden 47, 72, 74-5, 99-100, 123
Genii 116ff
Glastonbury 83
Golden Ass, The 118
Golden Legend 79ff
Gregory, Cedric E. 80ff
Guilds 133ff
Havilah 47, 72, 74-5, 99-100
Hephaestus 24ff, 34, 37, 39, 41, 84, 112, 128
Hercules 108, 109
Hermes Trismegistus 116ff, 121

Herodotus 44, 112
Hesiod 65
Hiram of Tyre 60
Hobbit, The 110-11
Holbein, Hans 81
Holdfast 110
Homer 21ff, 36, 40, 41, 43, 44
Homeric Hymn to Demeter 4
Iliad 22ff, 41, 43, 45
Incas 17
Jeans, Sir James 149
Jeremiah &Onias 78
Job 19
John, St 96
Josephus 56ff
Jubal &Jabal 48
Kabeiroi 112
Karma Yoga 121ff
Khnemu 45
Knapp, Daniel 82
Kreimhild 111
Lamb of Christ 74
Lamech 48-9
Leoben, Austria 81-2, 107
Logos 67
Los 142-3
Luther, Martin 129ff
Maat 45
Magi 72, 102
Marie Antoinette 148
Marsysas 36
Mathesius sermons 135
Memphis 44, 46
Menorah 55ff
Milton, John 115-16, 137
More, Thomas 6-7, 130
Moses 42-3, 50ff, 83, 86, 128
Mount Ararat 47
Mount Parnassus 47
Mount Sinai 42, 51ff, 54, 86, 128
Nebuchadnezzar 61ff
Nebuchadnezzar's dream 62ff
New Jerusalem 73ff
Nibelungenlied 111
Noah 47
Odyssey 21ff, 43
Olympus&Olympians 25, 36, 37, 77, 84

Oppenheimer, Robert 150-1
Ovid 3, 4, 35, 36
Palmer, Samuel 142
Pandora 44
Paul, St 30, 69-70, 78, 86-7
Psyche 118
Persephone 4, 100
Peruvian mines 17
Philo Judaeus 56ff
Plato 52ff, 65, 71, 74, 92, 94ff, 115, 122ff, 152
Pliny the Elder 5-6
Pluto 4
Prometheus 33-5
Prometheus Unbound 33-4
Pseudo Dionysius   see Dionysius the Areopagite
Ptah 44ff, 112
Puritans 135ff, 146
Red Ribbon Movement 18
Revelation, Book of 73ff, 94, 151-2
Robert of Chester 98
Scotus Erigena, John 90
Seker 45
Sekhet 45
Sephirothic tree 84
Shoemakers' Holiday, The 134
Siegfried 111ff
Slaves 14ff
Smaug 110-11
Snow White 118ff
Solomon 59ff
St Denis - Abbey 85ff
Stoics 152-3
Suger, Abbot of St Denis 76, 85ff
Tawney, R.H. 130ff
Temple of Solomon 59ff
Thebes 44
Thetis 24ff
Thor 76, 112
Thoth 45, 115
Tintern Abbey 83
Tolkien, J.R.R. 110-11
Tubalcain 48-50
Utopia 6-7, 130
Virgin Mary 101
Volsunga saga 111

Weber, Max 129ff
Wordsworth, William 10, 83, 138-9, 143ff
Xenophanes 153
Yoga 110-11, 121ff
Zeus 33-4, 36-7
Zimmern, Alfred 16-17, 19

www.ingramcontent.com/pod-product-compliance
Lightning Source LLC
Chambersburg PA
CBHW052004090426
42741CB00008B/1538